THE ART
AND
SCIENCE
OF
FRUGAL
INNOVATION

ADVANCE PRAISE FOR THE BOOK

'It is a very useful addition to the emerging literature on frugal innovation, frugal not just for the consumer and the manufacturer, but also for nature. I appreciate the authors' ability to differentiate between durable frugal innovations and the makeshift and jugaad ones'—Prof. Anil K. Gupta, CSIR Bhatnagar fellow, and founder, Honey Bee Network, SRISTI, GIAN and NIF

'This book makes a clear pitch for scientifically reasoned, simple yet affordable solutions that not only benefit the economically disadvantaged but also those who swear by sustainable development'—Prof. V. Ramgopal Rao, director, Indian Institute of Technology Delhi

'The world today is promoting innovation in every sphere of life to achieve the Sustainable Development Goals set by the United Nations. In this context, *The Art and Science of Frugal Innovation* is a unique book that has rekindled the concept and importance of frugal innovation in a lucid and comprehensive way, highlighting the science in frugal innovation and the value it creates'—Ravi K. Khetarpal, chair, Global Forum for Agricultural Research and Innovation, and executive secretary, APAARI, Bangkok

THE ART
AND
SCIENCE
OF
FRUGAL
INNOVATION

MALAVIKA DADLANI
ANIL WALI
KAUSHIK MUKERJEE

EBURY
PRESS

An imprint of Penguin Random House

EBURY PRESS

USA | Canada | UK | Ireland | Australia
New Zealand | India | South Africa | China

Ebury Press is part of the Penguin Random House group of companies
whose addresses can be found at global.penguinrandomhouse.com

Published by Penguin Random House India Pvt. Ltd
4th Floor, Capital Tower 1, MG Road,
Gurugram 122 002, Haryana, India

First published in Ebury Press by Penguin Random House India 2022

Copyright © Malavika Dadlani, Anil Wali and Kaushik Mukerjee 2022

10 9 8 7 6 5 4 3 2 1

ISBN 9780670093090

Typeset in Adobe Caslon Pro by Manipal Technologies Limited, Manipal
Printed at Replika Press Pvt. Ltd, India

www.penguin.co.in

Dedicated to all the imaginative minds who choose to address the myriad challenges by creating scientifically tempered frugal solutions.

Contents

Preface

Around a year ago, I had the privilege of meeting Malavika Dadlani in my office in Pune. We discussed the outline of a book that she was proposing to write on the role of science in frugal innovation. I strongly encouraged her to do that since I have always felt the need for India to move from jugaad to science-based systematic innovation.

I was, therefore, happy to know that Malavika, along with two co-authors, had completed the task of writing the book, appropriately titled *The Art and Science of Frugal Innovation*.

The book is very timely. The world is recovering from the coronavirus pandemic, which has caused all-round destruction. It has destroyed lives and livelihoods. It has destroyed economies. It has suddenly increased the inequalities, the disparities. At the last count, 49 million people had been pushed not just into poverty but extreme poverty.

So more than ever before, we need to make do with diminished resources and achieve deliveries of products and services to a large section of society. And that is where frugal innovation becomes important. That is where this splendid book by Malavika Dadlani, Anil Wali and Kaushik Mukerjee becomes very timely.

It was India that changed the dictionary of innovation. New phrases like 'frugal innovation', 'inclusive innovation', 'Gandhian innovation', 'indovation', 'more from less for more' and 'affordable excellence' have dominated the innovation dictionary, besides the prevalent grassroots innovation and jugaad innovation.

And in the last five years, there have been plethora of books on these subjects, with some of the books with titles with tweaking of the words, like better from less, etc. But all of them deal with frugal innovation one way or the other.

I am happy to say that this is the first book on frugal innovation that deals with its scientific underpinnings' emphasis on sustainable solutions.

The book nicely deals with diverse subjects ranging from market-driven solutions for frugal innovation to national policies and international conversations for inclusive innovation, ending finally by showing the way forward with sustenance.

Can we do frugal innovation with a strong underpinning of science? The answer is yes. Let me explain.

I started the Anjani Mashelkar Inclusive Innovation Award in my mother's name. This is the tenth year of the award. It is given for developing a technological solution that leads to inclusion, meaning that millions of resource-poor people can benefit from it.

One of these awards was given to a twenty-eight-year-old innovator, Myshkin Ingawale, the co-founder of Biosense. While discussing with one of his doctor friends, in Parol, Maharashtra, he realized that many women in villages were dying of anaemia because their low haemoglobin levels were not detected in time. He found out why: many of them were reluctant to give their blood for testing. So, he decided to create a non-invasive diagnostic tool, something that has never been achieved before.

He had to take recourse to high science. He used photoplethysmography, spectrophotometry and an advanced software for photon scattering to create ToucHb.

This was technological 'excellence' achieved by using cutting-edge science, and not jugaad. But it was affordable excellence.

Why? Because he reduced the cost per test from Rs 100 to just Rs 10.

In order that we move towards science-based frugal solutions, there needs to be a mindset change among our scientific institutions.

Our scientific community must understand that they can combine excellence and relevance. Indeed, they can achieve scientific excellence, thus publishing in high-quality scientific journals, as also relevance, by being relevant to the needs of the resource-poor society.

In 2014, Harvard University researchers created an inexpensive detector, costing just $25, that can be used by healthcare workers in the world's poorest areas to monitor diabetes, detect malaria, discover environmental pollutants and perform tests that are done by machines that cost thousands of dollars.

Work on this affordable $25 sensor was published in the August 2014 issue of the *Proceedings of the National Academy of Sciences of the United States of America*, one of the topmost journals in the world. The leader of this innovation team is George Whitesides, who, by the way, is the highest-cited scientist in the world with a staggering h-index[1] of 176! But he is dedicated to the cause of making high science work for the poor.

We need to create cost-conscious science and research. Currently, the frontier innovation is driven by expensive sophisticated research capabilities, which are popular among policymakers and the scientific community. The costs of high research and development (R & D) investments are recouped over a period of time through high-margin products. Whereas the driver for frugal innovation is the aspiration despite the scarcity of resources to create low-margin and affordable products. But high aspirations mean high-science-based and high-quality products, like what Myshkin did.

I hope this wonderful and timely book by Malavika, Anil and Kaushik will inspire the young generation of innovators to opt for high-science-based frugal innovation, which will bring prestige and pride to India for their magic of combining scientific excellence of a high order with relevance to large sections of our underprivileged society.

July 2020   Dr R.A. Mashelkar, FRS
                                                  Pune

Introduction

Frugal Innovations in Modern Society

Frugality has its roots both in Eastern and Western traditions. Major philosophies of the world, be it Buddhism, Epicureanism or Stoicism, religions like Jainism and religious texts like the Bhagavad Gita, all espouse frugality and simplicity in life, stressing upon the need for discipline and self-restraint. As early as in the eighteenth century, keeping in view the growing population and rising demands, philosophers like Adam Weishaupt stressed the need for frugality and moderation to maintain peace and social stability.[1] The philosophy of frugality helps remind us that resources are limited, so their use must be prudent.

The warp-weighted loom, an innovative weaving technology dating back to 7000 BCE, in which bundles of warp yarn were tied to hanging weights to keep the threads taut, is an example of need-based simple innovation that greatly increased weaving efficiency in ancient times.[2] Historian Samuel Noah Kramer recorded in 1963 that 'the people of Sumer had an unusual flair for technological invention'. The creativity of Sumerians (later known as Mesopotamians) in inventing new technologies and perfecting the large-scale use of existing ones, according to Philip Jones, associate curator and keeper of Penn Museum, Philadelphia, was driven to an extent by their lack of natural resources.[3] One can find many such examples of innovations due to the need for effective solutions using limited resources available to its people. Notwithstanding the unprecedented technological advancements that the world has experienced since the beginning of the last century, the stark inequality in distribution of wealth and resources among societies, coupled with concerns over sustainability of the earth, has encouraged the world to turn to frugality and conservation. In the world of novel products and processes, it's frugal innovations that are more likely to sustain and support inclusivity across economies.

The last decade has seen a surge of literature on the idea of frugal innovations, their significance, relevance, scope and limitations in today's world. The subject has mostly dealt with assessing their commercial utility, role in sustainable development and accessibility of technology. However, not only the context but the definition of frugal innovation itself has also undergone several changes during this period. In the beginning it was seen as a viable option for resource-constrained developing economies, as 'no-frills' and affordable innovations that could fulfil the needs of economically weaker sections and in the process also garner impressive profits by the sheer number of consumers at the 'bottom of the pyramid' (BOP), a term coined by the management guru C.K. Prahalad.[4] And now, frugal innovations are expected to provide better functionalities at lower cost, using resources more judiciously, turning waste to wealth, and promoting green manufacturing in a circular economy.[5]

In a world dominated by frugal innovations, affordability, superior performance and sustainability—and not mere premium pricing and abundance—will drive growth. Even a couple of years ago, an innovation that fulfilled basic features of a technology and was affordable to the BOP customers was considered a successful example of frugal innovation. However, in an era of fast-paced technological advancements, it is imperative that a new innovation not only gives an economical solution but also offers additional value over the existing options. In order to innovate for common customers, R & D goals need to be redefined. The research goals need to be targeted at developing technologically elegant yet affordable products, services and applications to meet the priorities, needs and aspirations of the vast majority of the people. It is increasingly becoming evident that such a goal cannot be achieved only by reorienting the manufacturing and marketing strategy but also needs a fresh look at scientific research and technology innovation, right from conception. For developing low-cost but high-level technological innovations, science and technology (S&T) institutions would need to adopt a philosophy of 'no frills, deep delving science' at the core of all research programmes and adhere to a sustainable path of technology development, embedding rapid technology turnarounds to gain from the indirect benefits of science to society. Firms also have to follow a new set of routines that are devised on the knowledge of price-conscious customers, and the markets dominated by low-cost products.[6]

Frugal innovations that offer high value at low costs are generally useful. Re-engineered solutions alone may not be good enough for a purpose. Low-cost launches do not always ensure success even when there is a latent demand for a product. The failure of such a product to take off can be attributed to many factors like low-value proposition, customer dissatisfaction, weak business model, product failure, etc., that can build a negative perception and impact its acceptance. The market is replete with instances where products even from leading companies have not survived: Tata Nano, Hero Honda Street, and Videocon's television are some examples. Thus, firms need to be cautious while planning to target mass markets through frugal innovations. In a highly globalized and connected world, perceptions, performance and consumer aspirations need to harmonize.

This book has tried to understand the hierarchy of resource-constrained innovations, which begins with cost innovations (same for less), good-enough innovations (tailored for less) and finally frugal innovations (new for less). Consider LOGIQ, a successful innovation of portable ultrasound device by General Electric (GE) Healthcare. It was not just 80 per cent cheaper than traditional ultrasound machines, but also smaller, lighter, offered core functions and was easily portable. More importantly, the R & D department of GE Healthcare continued innovating and adding new features in their subsequent upgraded models to fulfil the varied requirements of different markets. Frugal innovations are more challenging than good-enough stripped-down versions of products, and the success depends on marketing and product management capabilities.[7] With such products and services, it is essential to challenge conventions through social business models.

In this book, we have attempted to differentiate, but not separate, frugal innovations in terms of their stronger dependence on scientific interventions from all other types of low-cost innovations, commonly referred to as grassroots innovations, BOP innovations and jugaad innovations. Grassroots innovations are typically useful solutions for local communities and depend on the ingenuity of the innovators but the scope for expansion is often limited. Jugaad involves a make-do approach which does not involve the development of new technologies and the success of such innovations depends on 'flexibility' through recombination of resources. And finally, scalability is a challenge since the innovation is too localized and the diverse interests of various stakeholders have not been considered during development of the innovation. The overlap between frugal innovations and jugaad is in the 'inclusive' approach adopted for both innovations. We differentiate frugal innovations, as those requiring a 'clean-sheet' approach for various reasons. Unlike jugaad, a frugal innovation provides a durable solution. This approach has been referred to as the bricolage approach, wherein firms utilize their existing resource base to work out a solution. When a swift solution is found, the solution may not fulfil the tenets of frugal innovations. In some situations, jugaad innovations can be the starting point while a frugal solution would be the end point. Thus, the latter requires scientific planning, systematic testing and validation before commercialization. Even at grassroots, experimentation and testing are necessary before the solution can be adopted widely. Firms that aim to develop frugality-based advantage need to address external resource-based constraints or adopt more efficient design and development or business processes.[8] For instance, a majority of BOP customers own mobile phones, which can be leveraged by firms to promote technological solutions to users at the BOP. Firms like RML AgTech, formerly known as Reuters Market Light, and Nokia Life Tools Agriculture services send on farmers' mobile phones real-time information on crop prices. Digital Green sends out credible videos on innovative farming practices to groups of farmers. To create frugality-based advantage, it is essential to work around the constraints instead of making huge investments to bridge the gap in resources in emerging markets.

In this book, we have attempted to bring forth the role of scientific interventions at every stage of frugal innovation—from ideation to development and diffusion. In establishing our viewpoints, we have used some popular and some not so well-known examples from different fields, mostly from India, comprising medical services, diagnostics, vaccines and prosthetics, agriculture technologies, precision instrumentation, technologies towards better environment, sanitation, waste management and grassroots technologies, which were built upon innovative applications of well-founded scientific solutions.

The book proposes to present our views on frugal innovations from a different perspective and make people realize that science, an indispensable component of frugal innovations, is not practised in laboratories alone. As many proponents of frugal innovations believe, it is a philosophy, an alternative pathway to sustainable development, which can be successful only when innovators dare to act differently on what they feel is right.[9] It is time that scientists realized the relevance of frugality, and innovators and entrepreneurs used not just cheaper substitutes, but scientific knowledge and modern tools to provide solutions to the people at the base of the socio-economic pyramid. We believe that frugal innovations need to be empowering so that the technology is value-based, rooted in solid normative framings and socially controlled. As a result, inclusion of all forms of knowledge (e.g., indigenous, traditional or community knowledge) becomes appropriate from a socio-technical standpoint.

In general, innovations are inspired by situations. Extraordinary circumstances can play a big role in enabling rapid development of solutions to address problems. In the scenario of Covid-19, as the whole world grapples with the spectre of economic slump and societal challenges besides survival itself, frugal innovations are bringing hope. The situation gave rise to hundreds of start-ups making diagnostic kits, informatics tools, curative therapies, ventilators and protective gear, including facial masks, at a fraction of the prevailing prices for containing the spread of the novel coronavirus. Academic and scientific institutions accelerated the development of several innovative solutions and came out with outstanding results. From rapid, reliable and affordable diagnostics to preventive and curative treatments, from app-based healthcare to restructured mobile clinics, from 'Air Providing and Virus Destroying Mask' designed by a schoolgirl in India to the release of the first commercial mRNA vaccine by Pfizer and Moderna based on the pathbreaking research of Dr Katalin Karikó and Dr Drew Weissman of the University of Pennsylvania, innovative solutions at every level came from every corner of the world during the pandemic, which will be known for incubating a large number of potential innovations, many truly frugal. But, if there was one brilliant idea of frugal innovation that took shape during the pandemic, it was the conversion of empty railway coaches into Covid-19 hospital/isolation wards to manage a spike in hospitalization cases in India. It also granted the provision of moving to any hotspot within hours, making it invaluable. This is a perfect example of innovative and simple repurposing of existing facilities with minimal intervention.

The Council of Scientific and Industrial Research (CSIR) launched a Covid-19 initiative with five focus areas and developed twelve Health Assistive Devices (HADs) and five personal protective equipment (PPE) kits by September 2020 to fight Covid-19. This reaffirms our belief that India, with its scientific legacy and technological competence, the strength of its people and their wisdom, and wealth of traditional knowledge, can lead the world in frugal innovation. We hope this book will offer its readers a glimpse of some successful examples of frugal innovations and reflect on the unlimited possibilities of adopting S&T methods and tools to leverage human power and natural resources to create a better world.

1

Frugal Innovation—Emerging Perspectives

Engineering or technology is the making of things that did not previously exist, whereas science is the discovering of things that have long existed.

—David Billington

In April 2010, an article in *The Economist* described frugal innovation as 'products that are stripped down to their bare essentials taking the needs of poor consumers as the starting point'.[1] What this article did was to give a name to and conceptually articulate practices and approaches that have been in evidence for many years. For instance, efficiency, economy and elegance were attributed to the scholarly discourse of Prof. David Billington, who was a proponent of structural art at Princeton University, by his colleague Prof. Michael Littman, who said that 'efficiency means minimal materials, economy means minimal cost and elegance means maximum expression'. This would indeed be an apt description of frugality in a technical context. However, over the years, the scope of frugal innovation has expanded, covering products, business models and services.[2]

Frugal Innovation in Resource Scarce Situations

Of the many definitions of 'frugal innovation', the one given by Soni and Krishnan (2014) perhaps describes it comprehensively: 'a resource scarce solution (i.e., product, service, process, or business model) that is designed and implemented despite financial, technological, material or other resource constraints, whereby the final outcome is significantly cheaper than competitive offerings (if available) and is good enough to meet the basic needs of customers who would otherwise remain un(der) served'.[3]

The Covid-19 pandemic prompted the adoption of new approaches for frugal innovations optimizing the use of human capital and other resources. From managing the pandemic through lockdowns and enforcement of work-from-home (WFH) modalities, adopting digital technologies for daily needs to big data analysis and decision-making and controlling one of the most serious pandemics in living memory through rapid development of diagnostics, to drugs, vaccines and medical support system also reiterated the need for collaborative and open innovation scenarios.

Most frugal innovations originate while solving local and specific problems, using available skills and resources. The roles of science, technology experience, trial and error are extremely important, and innovators either adopt one or a combination of these approaches.

A seemingly simple innovation, which is built around a known technology or an existing practice, can sometimes change or expand the whole range of its application. For example, though many scientists contributed towards the development of integrated circuits (ICs), a set of interconnected electronic components that are integrated on a tiny chip of a semiconducting material, Jack St Clair Kilby, an electrical engineer from Texas Instruments, is credited with the development of the first hybrid IC in 1958 and its commercialization, which revolutionized the world of electronics. From audio, radio and optical devices to communication technology, medical instruments, implants, aircraft and spaceships, ICs are recognized as one of the most significant technological developments of the twenty-first century.[4] By bringing down the size, cost and energy consumption drastically and the manifold increase in the efficiency and precision of electronic devices, such innovations resulted in greater reach of technologies for commercial and non-commercial purposes.

In most cases, drastic cost reduction is achieved by using locally available resources, including trained human power, alternative processes, or by cutting down the frills and accessories that would not affect core functionality and efficiency. China has shown the power of home-grown technologies. As predicted by venture capitalist Rebecca Fannin,[5] China has established its tech supremacy in almost every sector with the power of frugal innovations, be it Tencent's WeChat messaging services, Haier home appliances or the Alibaba merchandising platform.

At the beginning of this decade, the primary purpose of all frugal innovations was affordable functionality. Innovations aimed at the BOP consumer were made to fulfil the needs and aspirations of a large segment of people having limited resources, Such frugal innovations were promoted on the merits of 'how to do more with less',[6] and were successful too. However, with all-round technological advancements taking place and rapid transformations happening worldwide, people's aspirations kept rising. As a result, expectations from frugal innovation are also expanding. Today's BOP consumers demand 'better from less', for which Radjou and Prabhu (2019) recommend six principles that promise a path towards sustainable growth.[7] Since such innovation is meant to be 'for all', these are broadly termed as 'frugal technology'. Depending on the complexity of problems and intricacies of innovations, these represent different levels of innovation (see chapter 2 for details) and known by many different names, such as reverse/inverse technology (innovations directed to achieve an end product by working it backwards), grassroots technology (innovations made and meant for the rural community), Gandhian innovations (a version adapted for lesser cost), nano-vation (a small but vital intervention), and of course jugaad ('quick-fixing' a problem using locally available resources and technology).[8] While any kind of frugal innovation (as mentioned above) is always cheaper than alternative options available in the market, at times these also offer a superior technology at a lower price.[9]

Though the term jugaad has a very Indian connotation of a local and quick-fix frugal, flexible approach to innovation, similar practices prevail in other countries as well, such as gambiarra in Brazil and jua kali in Kenya. Many grassroots innovations can also be seen as jugaad, which are of great value in certain situations and may or may not be replicable elsewhere. Though the notion of present-day frugal innovation initially originated from grassroots' innovations, mostly from India, China[10] and other developing economies, it soon encompassed a wide range of innovations, from rural electrification to medical diagnostics, with some based on deep scientific interventions and others redesigning existing technology to make it more affordable. These innovations, which may be called advanced frugal innovations, are low-cost sophisticated products made from minimal resource use.[11]

Therefore, frugality could mean different things in different contexts, but in the context of developing economies and resource use, whether it is drug development, medical management or building a small car, it must follow the principle of inclusivity.[12] It has been appropriately summarized as 'make more (performance) from less (resource) for more (people)',[13] which has now become a global mantra. A successful frugal innovation can be the result of scientific ingenuity in product development, or more efficient production, system of delivery or marketing strategy, which can transform an existing scientific or technological solution into a frugal innovation, as seen in the cases of Aravind Eye Hospital[14] in Tamil Nadu and Narayana Health[15] in Bengaluru.

Neither the cataract surgeries nor the cardiac procedures provided in these places respectively are new innovations *per se*, but by adopting a lean business model and several process and service innovations they have brought down the costs massively. India excels in service innovations applied to improve business models, delivery systems and institutional organization.

All scientific advancements do not necessarily lead to either frugal innovation or technology development. However, there is no denying that good science leads to good innovation. Scientific inventions and technology developments are the two wheels of the vehicle of progress on which a country rides towards the path of social, economic, political and overall leadership. Acknowledging that 'innovation is that third, and often the hidden, wheel which drives the vehicle of progress towards a sustainable goal', the United Nations Conference on Trade and Development (UNCTAD) Report, 2017, points out the need to adopt new modalities for development, bringing innovation to the foreground.[16] Innovative and improved technologies are key enablers of most Sustainable Development Goals (SDGs), whether it is achieving goal number two: 'end hunger, achieve food security and improved nutrition and promote sustainable agriculture', or fulfilling goal number nine:

'build resilient infrastructure, promote inclusive and sustainable industrialization and foster innovation'.

The question, therefore, is not whether to encourage frugal innovation, but what kind of innovation to encourage.

Does Frugal Innovation Need Scientific Intervention?

A study showed that consumers tend to prefer innovations that functionally improve their lives and yet are within their reach. Novelty is also an important criterion in the success of frugal innovations. But surely, consumers want value for their money. While a car that drives itself may be a truly disruptive innovation, more consumers are likely to buy waterproof smartphones or similar items of convenience.[17] On the other hand, though it would appear that a sunroof or waterproof umbrella cover—a frugal innovation to protect two wheelers from rain and the hot summer sun—would capture the market at an affordable price of Rs 500–600, it failed to survive because the consumers felt the plastic cover was too flimsy and got damaged easily due to wind force. Here the objective was well-identified, there was need for the product, but not enough experimentation was done to optimize the design, material and its functional value. Sometimes, frugal technologies are too simple to be accepted and users tend to doubt their sustained workability. In such situations, a systematic investigation is needed to validate the technology and gain users' confidence. Such is the story of Solar Water Disinfection (SODIS), a popular method, that works by exposing water to sunlight in a polyethylene terephthalate (PET) bottle. The method has been known since the 1980s but was not popular. Recognizing the potential of this innovation, which can be implemented without electricity and even under the most extreme conditions when potable water is scare, the Swiss Federal Institute of Aquatic Research and Technology started scientific investigations and multi-location field trials during the 1990s and implemented SODIS as an effective, simple and almost zero-cost water disinfection technology for drinking. Its studies established that solarization of water on a warm and sunny day in a clean PET bottle, including used and cleaned cola bottles of 500–600 ml, which allows passage of ultraviolet (UV) rays, can make water that is biologically-contaminated with bacteria, viruses, protozoa and worms safe to drink. The UV rays produce reactive oxygen species (ROS) from dissolved oxygen in the water, which, in combination with high temperature of about 50 degrees Celsius (inside the bottle) kill pathogens. About 6 hours of exposure on a sunny day was found sufficient and consumption of SODIS-treated water substantially brought down incidences of diarrhoea.[18]

Though the technology was much appreciated, concerns were raised about the possible harmful effects of the volatile by-products of PET released by the exposure to sunlight, which would be dissolved in the water. Therefore, systematic scientific evaluations were undertaken by the Centre for Affordable Water and Sanitation Technology (CAWST), Canada, and other scientists. It was found that very low amounts of adipates and phthalates are released in the water even after 17 hours of exposure at 60 degrees Celsius and are far below the acceptable World Health Organization (WHO) limits. Hence, the water is safe for human consumption.[19] These scientific validations paved way for adoption of this 'almost no-cost' method of water disinfection by more than a million families across Asia, Africa and Latin America at a negligible cost ($0.63 per person per year).

Some innovations ride on commercial gains, whereas others gain acceptance because of societal value. One such inspiring success story is of the Jaipur Foot, a prosthetic for the lower limb invented by Dr Pramod Karan Sethi, a professor of orthopaedics at Sawai Man Singh Medical College and Hospital, Jaipur, in 1968.

Concerned about the magnitude of the problem of disability due to the loss of the lower limb in India[20] and realizing that prosthetics available for lower limbs were either very expensive (costing more than Rs 60,000) or not user-friendly, Dr Sethi used his medical knowledge and the help of Ramchandra Sharma, a local artisan, to invent the Jaipur Foot. Designed to resemble a human foot, it was made of a waterproof and flexible material. Because of the flexibility, it allows foot-like dorsiflexion movements necessary for walking, climbing, squatting, cycling, etc. This affordable and versatile prosthetic was much preferred over the solid-ankle type prosthetic.

Jaipur Foot looks almost like the human foot. It has a heel block and a forefoot block, including toes made by gluing various layers of microcellular rubber, which is easily available. On the top, it has a wooden ankle block with a stainless-steel carriage bolt held in place by two iron nails, through which the rest of the leg can be attached. Several polymers are blended, rather than copolymerized, in the manufacturing of Jaipur Foot, thus making the process cost-effective, faster and much simpler.

The credit for scaling up this innovation and making it popular goes to a welfare society, the Bhagwan Mahaveer Viklang Sahayata Samiti (BMVSS), which engaged local artisans and used locally available good quality and high-density polyethylene (HDPE) for making the Jaipur Foot on a large scale. This innovation was aimed more at societal inclusion of people with physical disability rather than commercial gain, as BMVSS is committed to providing the services free of charge. Yet, it is the second-largest prosthetic foot provider in the world.

A major cause of concern with the Jaipur Foot was that it was heavier than other prosthetics and the manufacturing process required standardization. The team behind the Jaipur Foot was aware of this and, as their aim is to continuously improve the technology, intensive research was conducted for five years in collaboration with Indian Institute of Technology (IIT) Madras and National Chemical Laboratory (NCL) Pune to reduce the weight and standardize the fabrication materials and processes. The microcellular rubber used for production is basically polyurethane (PU) plastic. In spite of many years of research, a better substitute is yet to be found. However, such studies have helped improve standardization.[21]

The innovators received support from the Defence Research and Development Organisation (DRDO) in developing and testing the Jaipur Foot. The innovators continue to get technical support from the Massachusetts Institute of Technology (MIT) and Stanford University in the US; IITs Madras, Delhi and Bombay; and National Institute of Design (NID), Ahmedabad.[22] They have also been encouraged to devise a Jaipur Knee in collaboration with ReMotion Designs.[23] *Forbes* called the Jaipur Foot one of the most technologically advanced enterprises with high social relevance.[24] It is a prime example of a successful merger of scientific knowledge and innovation, leading to its commercialization. So, in this case, an excellent frugal innovation with high societal value could be bettered with more scientific interventions.

Engineering dropout Prashant Gade was shaken to realize that nearly 40,000 people lose upper limbs every year in India. Of these, the majority belong to rural areas, have meagre financial means and are forced to live with their disability as the available artificial upper arms are either not available or not affordable. He, thus, devoted himself to creating a robotic prosthetic arm to help rehabilitate such people. To achieve this, he dropped out of college, underwent specialized training in robotics and worked towards creating the Inali Arm, a robotic arm that has sensors to read signals from the brain, resulting in movement of the limb. It is India's most affordable robotic arm, costing only Rs 50,000, compared with prosthetic arms from China, the US and Europe, which cost Rs 10–12 lakh. He was recognized for his contribution by the Infosys Foundation, which awarded him the prestigious Aarohan Platinum Social Innovation Award, 2019.[25]

Frugal innovation often arises due to an impending problem, prompting the attention of the scientific community. Every year, India witnesses a large number of cases of dengue. The National Vector Borne Disease Control Programme (NVBDCP) found that in 2016, 1,29,166 people fell sick and 245 died from dengue. Treatment of dengue often gets delayed since it takes at least three–five days to diagnose the infectious disease. Navin Khanna, a member of International Centre for Genetic Engineering and Biotechnology (ICGEB) developed a kit called 'Dengue Day 1' that provides results within 15 minutes of the test and is also cheaper than other kits available in the Indian market. The kit helps determine whether the virus is in the primary or the secondary stage, and also detect all four types of dengue viruses, thus allowing doctors to treat patients more quickly. Although the kit developed by Dr Khanna is cheaper, more accurate and faster than other imported kits in the market, the Indian government was reluctant to buy it initially. During the 2013 dengue outbreak, imported kits ran out of stock, and fresh consignments failed to arrive in India in time. That compelled the government to purchase this indigenously developed kit, as a result of which the efficacy of the kit was acknowledged. Dr Khanna received the Anjani Mashelkar Inclusive Innovation Award for his valuable work. Acceptance of his technology and peer recognition encouraged Dr Khanna to continue working in this direction and develop a vaccine for dengue.[26] The scientific and organizational support of the ICGEB played an important part in promoting frugal science in its true spirit.

The need for organizational support and a strong scientific backing cannot be overemphasized in frugal innovations. While Dr Khanna, a regular research staff of ICGEB, had the advantage of its scientific support, independent innovators have to seek out research institutions in India to receive this kind of help. Thankfully, there are many instances of such support being extended, which will be covered in later chapters.

There are numerous start-ups that take up low-cost and easy-to-access technologies to serve a larger section of population living below the poverty line. They utilize technological advancements in artificial intelligence (AI), biotechnological tools, information technology (IT), block chain and big data analytics as well as engineering tools and devices to substitute human labour at various levels. The single common factor underlying such programmes is to devise solutions that are cheap and easy to penetrate markets far and wide within a short span of time and reap maximum profits by large volumes. Though many receive support from academia and corporate R & D in giving shape to their innovative ideas, there are others who work on their own, using only the facilities available to them at the place of their study/work, or at incubation centres.

Achira Labs, founded in 2009 in Bengaluru, is a tech company that builds analytical medical diagnostic platforms, lab-on-chip, using the microfluidics approach. Microfluidics is a set of technologies or tools that can be used to manipulate fluids at a sub-millimetre level, which not only miniaturizes operations, but also allows integration and automation of many functions. Dr Dhananjaya Dendukuri, one of the co-founders and CEO of Achira Labs, first worked on this concept during his PhD at MIT. After returning to India, Dr Dendukuri started Achira Labs to turn his concepts into viable technology.

Achira's lab-on-chip consists of a microfluidics chip and a reader that can diagnose disorders associated with thyroid or female infertility. Tests linked to these disorders can be easily multiplexed, allowing multiple parameters to be tested using a minute amount of a patient's blood or urine sample. 'The idea is to remove testing from a centralized paradigm and move it to the doctor's office. The sample, along with the buffer, would be loaded on the chip by a technician, which would then be fed into the reader. The microfluidic chip, which is smaller than a credit card, would have channels etched on them, through which the sample will flow. This platform can rapidly (within 30 minutes) be used for an array of tests with much reduced use of reagents and high sensitivity, eliminating the use of large analysers. This not only makes the process easier, but also cuts down the cost.'[27]

The examples given above clearly show the diversity of purpose and varying needs for scientific interventions in developing frugal innovations, though each was based on a scientific concept.

Early Proponents

Frugal science is not new to us. Many simple devices of everyday use are the result of it. In pre-Independence India, the scientific community had limited resources and means to pursue research in the field of their choice but was being exposed to the revolutionary technological advancements being made by the Western world in every field of science. The desire to prove their scientific competence, even with limited resources and frugal substitutes, ignited the innovative spark among many Indian researchers.

The work of Sir Jagadish Chandra Bose, the first modern scientist of India, is a brilliant example of frugal science. A man ahead of his time, he invented the crescograph, 60 gigahertz (GHz) microwave apparatus, galena crystal detector, horn antenna and many other items based on sound scientific grounds, using locally available and cheap resources to demonstrate some world-class scientific experiments.

In the years after Independence, with better availability and access to resources, India paid attention to modern ways of scientific research. International competitiveness and dominance of market-driven technology shifted the focus from frugal science to upstream research and experimentation, which necessitated the use of expensive machinery, equipment, tools and high-grade chemicals for the better part of the last millennium.

However, towards the later part of the last century, the world had started experiencing the many fallouts of rapid industrial growth, with diminishing non-renewable resources in the face of a burgeoning world population and climate change, and their inevitable impact on world economy widening the inequality. These changes compelled the world to look at things differently, and to innovate and seek affordable solutions to meet these challenges. The most direct impact of climate uncertainties is seen in agriculture productivity and production, with a meltdown effect, particularly in agrarian economies like India. As a consequence, conservation of resources and inclusiveness in innovation became imperative and the world is once again focusing more on frugal science, leading to frugal innovation, to make most from the least.

Contrary to the general belief that frugal science means simplistic science, requiring a lesser degree of scientific competence, it takes a much deeper scientific understanding and foresight to come up with a frugal technology of high acceptance. Unlike jugaad, science-based frugal innovations are hardly deployed to quick-fix a problem at hand. These are technologies based on scientific grounds and are the results of a strategy to develop products or services for a limited resource ecosystem. Such technologies can be truly disruptive because of their deep penetration across economies.

Revival of Frugal Science

It is encouraging to find that many science leaders today are not only practising frugal science themselves but also spreading the philosophy by mentoring young minds to think out of the box. Whether it is Prof. Vijayraghavan Chariar of the Centre of Rural Development and Technology at IIT Delhi or Dr Manu Prakash of Stanford University, they think frugal, work on a frugal budget and build sophisticated but affordable technology that more and more people can use. Dr Prakash presents a brilliant case advocating frugal science over conventional practices.[28] Dr Prakash believes that more than accumulating data and depending on information, one should lay stress on 'experiencing the concept and taste the science'.[29] For scientists and academicians like Dr Prakash, frugal science is that link between knowledge and experience. Today, Dr Prakash's inexpensive inventions let everyone—children, historians, medical diagnostic technicians and farmers in every part of the world—learn about science by experimenting using affordable tools that do not need an expensive science laboratory, and function just as well under a tree in a village. Dr Prakash believes that scientific tools need to be made more accessible and cheaper, and 'it's important to bring open-ended tools for discovery to a broad spectrum of users without dumbing down the tools'. To achieve this, Dr Prakash encourages interdisciplinary research to connect physics and material sciences to biology, medicine and environment and advocates frugal science to develop low-cost tools for health care, teaching and environmental protection. He has inspired a whole band of young researchers and practitioners of science, not only in Stanford University, but by example, in many other institutions in the US and elsewhere. When faced with non-accessibility of a vital instrument, he has turned to make functional substitutes rather than complain, beg or bypass its use. In the face of one such challenge, this assistant professor of bioengineering devised a microscope called the Foldscope, which is made mostly of paper and costs around Rs 50 (less than $1). It replaces the precision-engineered lens housing that is a traditional microscope's costliest component, with a paper assembly like an old-fashioned slide rule. The pattern is printed, cut and folded from a single sheet following the precision method as in origami. Using inexpensive lenses, the Foldscope can magnify 140 times, enough for significant healthcare and educational applications.[30]

Clinics in Tanzania and Ghana have used the Foldscope to diagnose schistosomiasis and malaria. Thousands of children around the globe have made their own Foldscopes to learn science. Dr Prakash often combines the working principles of old tools, instruments and even toys in developing versatile and robust new scientific devices. For example, along with graduate student George Korir, Dr Prakash developed a hand-cranked chemistry tool that uses nineteenth-century paper punched tape technology to power a microfluidics chip. Unlike comparable devices, this $5-tool (about Rs 300) doesn't need electricity or batteries and so it is ideal for tasks like testing water quality in a remote village.

As Dr Prakash sums it up, 'the novelty of frugal science lies in looking at things differently. Sometimes what we are searching is right in front of us, we only need to change our perception to "see" it'.[31] Dr Prakash's inventions are being used by the global community, from Nigeria to India. Supported by the Department of Biotechnology (DBT), Ministry of Science and Technology, Government of India, Dr Prakash and his team have held workshops in India to propagate the use of Foldscopes and other laboratory tools. The Prakash Lab team has distributed thousands of Foldscopes in 135 countries.

The beauty of frugal science lies in finding incredible science hidden in ordinary things and mundane acts of everyday life.[32] It could even be a simple whirligig, as it was in case of Saad Bhamla, Dr Prakash's former colleague from Stanford University and another advocate of frugal science.

Dr Bhamla prefers to make technologies based on frugal science that are free for use. Innovator of laboratory devices like the Paperfuge and the Electropen, Dr Bhamla, a chemical engineering graduate from IIT Madras, worked on the problem of 'the dynamics of the tear film in the presence of contact lenses' for his doctoral degree at Stanford University. A polymath like Bose, Dr Bhamla has also forayed into various areas of science without being limited by the boundaries of specialization.

Recognizing that laboratory centrifuges, an essential tool in almost every study in biology, medicine and material sciences, are expensive, heavy and require electricity, which makes it difficult for the developing world to perform daily routine tests, he developed a hand-powered ultra-low-cost paper centrifuge— the Paperfuge. This lightweight (2 gram) and inexpensive ($0.20 or Rs 1.30) human-powered paper centrifuge was designed on the basis of a theoretical model inspired by the fundamental mechanics of an ancient whirligig, a buzzer toy, from 3300 BC. This simple toy, which many of us might have played with at some time, has a circular disc that spins by pulling the strings passing through its centre in opposite directions. It is amazing that such a device, by simply using hand power can achieve speeds of 1,25,000 revolutions per minute (rpm), equivalent to 30,000 g of centrifugal force, sufficient to separate pure plasma from whole blood in less than 1.5 minutes! Centrifugal microfluidic devices like the Paperfuge can isolate malarial parasites in 15 minutes and hence promise immense application in diagnostics in resource-poor situations.[33] Like Dr Prakash's Foldscope, the Paperfuge also has a huge scope for use in teaching at the school and college levels. Not only will it be a cost-effective substitute for a centrifuge, which is an essential instrument required to teach science even at the school level, but by involving students in making this device, the complex theories of physics relating to motion can be made very simple.

Dr Bhamla, while working on a synthetic biology project, also motivated and guided high school students to create a small (12 gram), portable and very cheap ($0.20) electroporator (Electropen) made with a 3-D printer, aluminium foil and a cigarette lighter. This Electropen was found to be comparable to an industrial grade electroporator, which can be used in molecular biology experiments. The simplicity and affordability of such devices help make science experimentations accessible and their application global. More importantly, these innovations set exciting examples for their followers, prompting them to be consumed by the sheer elegance of scientific innovation. Prof. Anil Gupta, a strong believer of frugal innovations, and founder of the National Innovation Foundation (NIF), Honey Bee Network, and Society for Research and Initiative for Sustainable Technologies and Institutions (SRISTI), believes that every young science researcher in India must be encouraged by the faculty to explore scientific wealth in indigenous knowledge, amalgamate formal and informal science, and convert these into frugal innovations.[34]

The interest in science for the sake of science alone and desire to innovate something that has wider social relevance and use are common to both Dr Prakash and Dr Bhamla's work. It is no coincidence that both Dr Prakash and Dr Bhamla worked at Stanford University, which is known to encourage innovative research, and that Dr Prakash also helped in designing Dr Bhamla's Paperfuge. Clearly, organizations, peer groups and mentors play an important role in shaping one's ideology and goals. On their part, believers of frugal science not only practice it in their scientific endeavours, but also inspire and encourage others to do the same.

Saving natural resources, protecting the environment, and providing access to the poorest of the poor are the core principles of frugal science. However, unlike frugal innovation in a product of commercial value, a highly acclaimed frugal scientific innovation may not have big market potential. Being innovative and at the same time having links with potential users and mass manufacturers are essential requirements to make frugal innovation successful, whereas frugal science remains relevant even when it is not commercially successful. However, it would be a mistake to think that R & D investment on a frugal innovation project is always stringent. Often, large expenditure is needed for undertaking research that leads to the development of a frugal solution. A handheld, low-cost and smartphone-connected full-body ultrasound-scanning device was developed by Butterfly Network, US, thanks to enormous funding of $300–350 million by Fosun Pharma, Bill & Melinda Gates Foundation, Jamie Dinan and many other investors. However, such expenditure seems very much justified, considering that the product is to be sold for $2,399, against traditional ultrasound scanners (USS) that cost $50,000–$100,000. The scientists at University of British Columbia, Canada, have developed wearable USS consisting of tiny vibrating drums that are made of polymer resin, called polymer capacitive micro-machined ultrasound transducers (polyCMUTs), that are considerably cheaper to manufacture than the available standard scanners.[35] They are working to bring down the cost and to develop a wearable USS at $100, which can be used as a diagnostic tool for millions in the world. The new transducer, once commercialized, can revolutionize the use of USS for numerous diagnostic purposes, especially in areas with scant resources and limited access to modern medical facilities.

There are many examples where a leader, mentor or colleague has deeply influenced an individual or the whole team or organization to follow the path of frugal science and development of world-class technology. Beginning her research career in a modest laboratory of Calcutta University under the mentorship of Prof. R.N. Basu, a strong believer of the philosophy of frugal science, Malavika imbibed the importance of frugality in science at an early stage. Funds for research were limited but the quest for science was deep. So the practice was to extract more from less by using local and cheap substitutes wherever possible. For the first time she learnt a unique way of using discarded glass containers for laboratory work. These would be purchased at throwaway prices from local scrap dealers and put through a series of washing, cleaning and sterilization processes. These were then fit to be used in biological experiments in place of expensive laboratory glassware. Hundreds of combinations of nutrients and growth substances were tested in different concentrations to find the best ones that would promote the rooting of cuttings in different 'hard to root' plant species. This was her first lesson in practising jugaad in scientific research without compromising on quality. This led her and her teammates to devise a simple and improved technique to study seed germination. It used twelve to eighteen square glass plates of 8–9 inches, placed in a standard plastic slanted dish drying stand, that one could buy in the 1970s for Rs 20 at any regular store. Seeds were arranged over the glass plate over blotter paper and germination was recorded without disturbing the seeds. The cost of setting an experiment using twelve plates cost less than one-tenth of the regular practice of using glass Petri dishes. This indeed was an innovation that was technically superior, cost-effective, most innovative and locally accessible.

But the real meaning of 'frugal science' was understood after years of dedicated team research under the leadership of Prof. Basu. A number of almost 'zero-cost' wet and dry protocols for seed treatments were standardized. These improved germination and shelf life of seeds. The technology was effective for a large number of plant species such as wheat, rice, jute, lentil, sunflower, cotton to carrot, lettuce, onion, etc.[36] The technology was frugal, simple, effective and used locally available substances for seed treatment, ranging from simple water to common salt, bleaching powder, over-the-counter aspirin tablets, arappu (*Albizia amara*) and tulsi (*Ocimum sanctum*) leaves, and many other substances of biological or chemical origin. These protocols not only proved to be effective in prolonging the shelf life of seeds, but also presented a new scientific hypothesis explaining the mechanism of seed ageing.[37] Being simple, inexpensive and effective, this on-farm innovation was widely adopted by farmers and also commercialized as low-cost dry-seed treatments, mostly in the eastern and southern parts of the country,[38] where the situation is more unfavourable due to the prevailing humidity.

It is clear that in all these cases, innovative ideas took birth due to the keen observations of scientists, followed by experimentation and development of 'proof of concept'. These were then given sound scientific support and validated by the multi-disciplinary teams of their respective universities. Many similar examples can be drawn from research institutes in India and other countries.

The best example of a scientific institution in India where cutting-edge science, frugality and innovativeness are the guiding principles is the Indian Space Research Organisation (ISRO).

An Organization Guided By the Principles of Frugal Science

Headquartered at Bengaluru, ISRO was established in 1969, with the vision 'to harness space technology for national development, while pursuing space science research and planetary exploration'. It was passionately led by visionaries like Dr Vikram Sarabhai, Dr Satish Dhawan, Dr U.R. Rao, Dr K. Kasturirangan and others. Though the space programme itself may not be classified as frugal science, what scientists like Dr A.P.J. Abdul Kalam, the project director of India's first Satellite Launch Vehicle (SLV-III), believed in and promoted were examples of frugal science. Scientists work out strategies to make the most of the least with the best performance, which are nothing but examples of grit, ingenuity and frugal science.[39]

It was the vision of the leaders and policy planners of the time to justify such huge expenditure into something that was not considered amongst the top priorities of the nation. As rightly pointed out by Dr Rajeshwari Pillai Rajagopalan (2019), the head of the Nuclear and Space Policy Initiative, Observer Research Foundation, 'to justify spending precious resources, India's space programme has focused on developmental missions right from the beginning; mainly establishing communication satellites, weather forecasting, and remote sensing technology. It has since become one of the most cost-effective space industries in the world'.[40] On 22 July 2019, when the second mission to the moon, Chandrayaan-2, was successfully launched from Sriharikota, the country celebrated this scientific feat like never before.

India's space research programme has earned global recognition not only for using indigenous technology that is world-class, but also for achieving this at a fraction of the cost of similar projects in other developed countries. For example, the cost of the Mars Orbiter Mission (MOM), or as it is known popularly, Mangalyaan, was $74 million, which is roughly ten times cheaper than National Aeronautics and Space Administration's (NASA's) Mars Atmosphere and Volatile Evolution mission (MAVEN) orbiter, which cost around $671 million. Several countries are now looking to India's space programme to launch their satellites and utilize ISRO's expertise in developing their own satellites. In 2017, the US, Israel, Kazakhstan, the Netherlands, Switzerland and the United Arab Emirates used ISRO's single rocket, the Polar Satellite Launch Vehicle (PSLV)-C37 to launch 104 satellites, creating a world record.

India has been able to keep the cost of its space programmes low due to several factors, such as keeping the programmes simpler than organizations like NASA, leveraging the availability of scientific competence within the country, and of course cheap human workforce. But above all, it was the will and the leadership that made the difference. There is tremendous opportunity for the Indian space programme to become the preferred partner of many south American and African countries keen to establish their space programme at an affordable cost. According to a report in the *Financial Express*,[41] the global space market is poised to expand in South America and Africa, and ISRO's commercial wing Antrix Corporation Ltd is expected to handle big business in the years to come. There will also be business opportunities for ancillary industries and scope for IT professionals, where India has an edge. To foster a viable public-private partnership in the space sector, the Indian government has created a new commercial enterprise, New Space India Limited (NSIL).

Frugal Science—A Philosophy

Frugal science derives from self-sufficiency and is not dependent on expensive equipment, tools, diagnostics and technologies. It advocates an analytical and innovative temper that nurtures out-of-the-box solutions based on robust scientific concepts. Hence, developing effective and affordable technological solutions or technologies for further scientific deliberations with limited use of materials, machines, energy and human power has become the central idea of frugal science, even though its role in social reorganization may not be evident immediately. Therefore, we propose that the scope of frugal science can be widened by defining it as a science-based approach to create affordable devices, tools and techniques or a verifiable solution to a problem of great challenge that can replace or better the existing technologies or provide a new solution to a large population at a substantially reduced cost.

In the realm of social sciences, frugality represents an alternative path for sustainable development.[42] Even in technology development, frugality is becoming a key criterion for sustainable application, though it is generally believed that frugal and other similar innovations (namely, jugaad, resource-constrained innovation, inclusive innovation, BOP innovation, etc.) are terms for flexible and affordable solutions, that are based on the fundamental assumption that poverty is caused by the lack of resources.[43]

The appreciation for frugal science comes by working in an environment where the majority believes in frugality, both in use of resources and the outcome, without compromising on quality. But, in most situations, it is the commitment of leadership towards frugal approach that carries the day.

Frugal science generally stems from frugal behaviour. At least with a large number of people in countries like India, such a propensity can be ascribed to cultural upbringing, specific socio-economic circumstances or even spiritual moorings. The inclusion of frugality and an austere approach in activities such as institutional R & D is a welcome addition. To address people's burgeoning needs, from energy to materials, we have no choice but to move away from solipsistic science[44] and embrace a new brand of frugal science—one that holds promise to usher in both sanity and sustainability in our lives.

2

Science in Frugal Innovation

I would prize every invention of science made for the benefit of all.

—Mahatma Gandhi

Scientific temper is a prerequisite to innovate. An attitude that involves the application of logic, argument and analysis is a vital part of scientific temper. Being innovative is the key to problem solving, sustainable development and inclusive social transformation. Good frugal science is always a harmonious combination of innovation and scientific temper. Science is indispensable not only for understanding our existence and relationship with our surroundings, but also in addressing the major challenges in our lives and finding viable solutions to manage them.

Science cannot just be limited to systematic studies based on experimentation, which lead to newer explanations, hypotheses, inventions and innovation; it is also a way of looking at things and finding verifiable explanations to events and situations around us. True to the observation of Margaret Wheatley, a profound thinker, speaker, and author of nine books on various aspects of human behaviour,[1] living systems use their knowledge to tide over challenges at multiple levels on different timescales. Scientific knowledge and analytical ability help us understand changes and find solutions to ever-emerging challenges by learning from existing knowledge and through new experimentations, which are seldom confined to research laboratories. Elements of fairness, prudence and frugality in resource use, which are built into scientific temperament, are also at the core of frugal innovation. Hence, the famous saying of American physicist Brian Greene, 'Science is a way of life. Science is a perspective', aptly describes the role science plays in every innovation.

Generation of scientific knowledge is a continuous and cumulative process. Hence, what we know today is either built gradually in a manner of progressive explanations or is the result of a disruptive innovation discarding any pre-existing hypotheses. By using a scientific concept or a basic technological innovation, several divergent technologies can be built to meet the newer challenges. Science can help expand an idea into an innovation and convert a single innovation into several useful applications.

If scientific inventions marked the last three centuries and took the world through industrial revolution and growth, the present millennium will be remembered for innovation in every walk of life. The relationship between scientific knowledge, technology development and innovative applications are interdependent and multidirectional, as shown below. For the development of an efficient solution at an affordable cost, a technology must be innovative, scientifically sound and must have utilized minimal resources. However, an important consideration, in case of frugal technology is sustainability, which also requires continuous scientific validations, necessary interventions and modification of technology from time to time, failing which, the innovation may not last long.

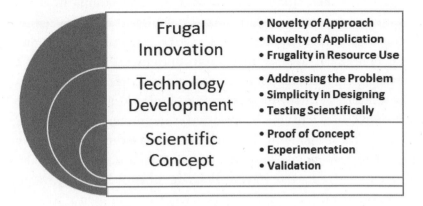

Translating scientific knowledge into a successful technology very often depends on the innovative approach in which it is applied. Therefore, science, technology and innovation are highly interdependent when it comes to application in life. A simple innovation, applied to an existing concept or technology, and utilizing local resources, can sometimes change the entire scope of its application by making the technology affordable and more user-friendly. By bringing down the cost of a process or product, such innovations result in greater deployment of technologies for commercial or non-commercial purposes. Emergence of user-friendly applications for smartphones, and mobile-based cardiac monitoring devices are two such examples of successful Frugal Technology in the past few years.

The idea of an innovation may come from common knowledge, traditional practice, borrowing from an existing but divergent example or methodical research. But in most cases, a sound scientific principle is at the core of an innovation. In others, scientific knowledge plays an important role in fine tuning and upscaling the innovation into a successful product or service.

Physicist Sir C.V. Raman, while returning from his first overseas trip to England in 1921, was perplexed after observing that sea water looked brilliant blue in colour but the same water when put in a glass looked colourless. Not satisfied by the explanations available in scientific literature at the time, he conducted a series of simple and frugal but systematic experiments in collaboration with his students and colleagues to find a scientific explanation as to how a light beam passing through different media changed colour depending on the molecular structure of the substance. This led to the discovery of the phenomenon known today as the Raman effect, which earned India its first Nobel Prize in physics in 1930. Based on this principle, Raman spectroscopy is commonly used in chemistry to provide a structural fingerprint for identifying molecules and an array of related purposes.

An innovative device, the Raman smart needle, was recently developed by a team led by Prof. Nicholas Stone of UK-based University of Exeter by applying the principles of the Raman effect. This has brought forth the fact that even after almost a century of Sir Raman's observation and the subsequent preposition of the Raman effect to explain the inelastic scattering of light through different substances (molecules), it not only remains relevant but is also being applied in many innovative ways to develop important devices.[2]

Pisoni (et al., 2018) observed that the real challenge for frugal innovation is to introduce something new or different (innovating), whilst being frugal with resources.[3] Given the fact that scientific research is an expensive but essential instrument for the progress of any country, research organizations strive to innovate not only to find better solutions to prevailing problems that help increase human efficiency and reduce drudgery at an affordable cost, but also to create value and make a profitable business. From identifying the priority areas to crystallization of an idea and conceptualization, planning and execution of the programme, a series of systematic activities of scientific experimentation, analysis and validation are to be undertaken for a successful outcome.

The relevance of frugal innovation is often seen in the context of scarcity and conservation of resources at one end and affordability at the bottom of the pyramid on the other.[4] However, without strong scientific backing, a solution, in spite of being frugal, stands a slim chance of being impactful. Interestingly, the opportunity for India to emerge as the 'frugal innovation hub' is immense as it has the world's second largest number of S&T professionals and a congenial socio-politico ecosystem. Handled properly, this can be a game changer, fitting well with the theme of 'make in India and sell globally'.

The world has come to realize the importance of frugality in the success of an innovation. The fierce competitiveness between manufacturers and service providers, and shortening of product life and hence, market size due to changing preferences of consumers and fast replacement of products has brought forth the role of science as the key driver. The right combination of science, technology, and management are crucial to ensure success with frugal innovations. Business managers and technologists need to work in tandem to help fulfil consumers' aspirations through innovative frugal solutions by leveraging the ingenuity of scientists and innovators. To understand the phenomenon of frugal innovation, researchers have attempted to disaggregate it into a basic mindset or philosophy of frugality; frugality in the process or workflow and its manifestation in a product or service.[5] In a nutshell, frugal science is the engine that drives a vast range of frugal innovations, fuelled by the imagination of the creator, aligned on the track by appropriate strategies for upscaling and commercialization and guided by national and international policies and regulations.

Scientific Excellence Shines through Frugality

The Bose Institute, one of the CSIR institutions, was established in 1917 by Sir Bose, one of the greatest scientists of modern India. It is popularly called the Basu Vigyan Mandir, i.e., Bose's Temple of Science. And a temple it is. Seeing replicas and prototypes of remarkable and groundbreaking inventions by one of India's most innovative physicists-turned-plant-physiologists, Sir Bose, one feels humbled and awestruck by his ingenuity and scientific brilliance. He contributed significantly in spite of meagre resources at his disposal. It was a rare combination of scientific genius, and innovativeness in experimentation and designing of instruments with locally available and affordable materials that established his undisputed scientific credentials. He also innovated several scientific devices, small and big, such as the iron–mercury–iron coherer with telephone detector, carbon-coated rotatory recorder and many now-commonplace microwave components, which served to validate new scientific hypotheses. His work was a prime example of frugal science, though the term was yet to be coined.

Frugality in scientific research or innovation is not just a consequence of having limited resources. It is a way of thinking, an inclusive mindset that does not justify the needless use of expensive tools, instruments, machines, chemicals and other resources. The idea is also to always work towards producing an affordable end product. The shortcomings and imperfections, if any of a frugal innovation can be minimized by a strong scientific backstopping. Scientific validations also help build user confidence and hence, better adoption of an innovation.

The story of Dr Hemendra Nath Chatterjee, a medical practitioner from Kolkata, could have been different, if his innovation had been backed by sufficient scientific data. Dr Chatterjee pioneered a treatment for cholera in the early 1950s by administering an oral glucose-sodium chloride solution to rehydrate patients, which saved hundreds suffering from mild to moderate cholera. Though he published the results with two co-authors in the *Lancet*, a reputed medical journal, lack of maintaining necessary checks (control) and supporting data on fluid balance, etc., raised several serious questions about its scientific veracity. This simple and innovative therapy was a significant improvement over the then-prevailing intravenous administration of saline glucose to treat cholera. It was a truly frugal innovation, and was readily adapted by doctors in many Asian and African countries.[6] However, it took almost twenty years of supplementary data generation by scientists like Dr Dhiman Barua and Dr Dilip Mahalanabis, before the oral rehydration salts (ORS) and oral rehydration therapy (ORT) were internationally recognized and accepted by the WHO.[7] Thus, neither the scientific merit nor the philosophy of frugality behind the innovation of ORS could be appreciated because of the lack of its scientific validation.

Nearly every technical professional, be it a medical practitioner, a pharmacist, agriculturist, engineer, architect or computer programmer, there is an intense desire to create something new one day. Almost every S&T professional dreams of being identified by at least one innovation, small or big, that will be recognized by one's peer group. But only a few succeed. It is rare that the innovation made is universal in application and accepted across disciplines. Generally, a path-breaking innovation, frugal or not, happens when least expected. On the other hand, in our daily lives, we spontaneously innovate many things—products as well as processes. There are innumerable examples of such creative use of knowledge to devise an innovative application, from a new recipe to a modified household appliance or developing an innovative learning model for teaching alphabets to a child. An innovative housewife or an intelligent teenager can do these so effortlessly that we do not even recognize them as innovation. Science, therefore, is more about logical deductions and a methodical approach in life than being limited to complicated experiments in the confines of laboratories. And many of these innovations are frugal by nature, not by design.

Problem Solving at the Core of Frugal Innovation

Given their circumstances, resource-poor farmers and artisans often devise innovative techniques to increase productivity, using fewer resources to develop new and more efficient devices or farming technologies. Shrawan Kumar Bajya, a farmer from the village of Girdharipura in Sikar district of Rajasthan, made an onion harvester, a salt turner and a multipurpose weeder in less than half the cost of the standard available machines. He used the engine of a motorcycle and spare parts from other two-wheelers or attached accessories like an onion digger, a weeder or a salt turner to a motorcycle. These innovations not only won him a President's Award in 2017 [8] and the NIF's Grassroots Innovation Award,[9] but are also helping hundreds of his fellow farmers. Bajya isn't a scientist but he thinks scientifically and innovates to solve challenges he faces in his life.

Consider the weavers of Telangana, Andhra Pradesh, Odisha and Varanasi, who are making new patterns by combining traditional knowledge with computer-aided design (CAD). This not only maximizes their efficiency but also expands their product range. These examples clearly show that being creative and innovative is natural to human intellect. However, converting ideas into innovations for mass use sometimes necessitates inputs from a technical expert, or organizational support in scaling up and commercializing.

The story of Mitticool, a refrigerator made of clay that does not require electricity, is quoted often in many articles on frugal innovation. Encouraged by the success of 'Mitticool', its creator Mansukhbhai Prajapati, a potter from Wankaner, Gujarat, worked to commercialize clay cookware, including pressure cookers, water filters and a range of crockery, which created great business opportunities for the other potters of Gujarat and him. The Honey Bee Network and the NIF played the role of enablers and technical-support providers. In fact, it is also an example of collaborative innovation networks involving the product and process capabilities, which are well recognized [10] at improving innovative product performance.

One might classify many of these as jugaad innovations or frugal innovations based on the Doing, Using and Innovating or the D-U-I process,[11] which does not apply systematic and scientific planning, experimenting or validation. Since most of the above cases dealt with finding solutions to situation-specific problems, and barring Prajapati's innovation, did not aim at a large diffusion, even a jugaad strategy with limited spread of the technology was good enough for localized users.

The Success of Frugal Innovations

A distinguished faculty of Harvard University for over 50 years and an intellectual leader in the field of science and public policy, Harvey Brooks (1915–2004), while analysing the interdependence of science, technology and innovation, each of which represents a distinct and large category of activities, identified at least six ways in which science contributes to technology. The most important of these was the creation of a knowledge base that enables more efficient strategies of applied research, development, and refinement of new technologies.[12]

Every now and then, we learn of a product that may reduce air pollution, an app that can detect heart attacks or a method to turn polythene waste into alternative fuel for cement industries. Each one promises to solve a critical problem faced by people in their everyday life, or even have a larger impact on the environment. While many of these get media coverage and even receive academic acclaim, only some achieve commercial success by becoming a part of peoples' lives.

The roadblocks to success may be due to any of the four main stages in the cycle of innovation. Starting from ideation, these are testing for proof of concept (PoC); late-stage product development, and finally commercialization in the marketplace (Renu Swarup, 2018). The role of scientific validation becomes most crucial at the first three stages, whereas the role of market strategies is more vital in creating awareness, generating demand, assessing the sustainability of business and aligning with the regulations, legality and business practices. This was well demonstrated by Dr Susanta K. Roy's frugal innovation, the zero-energy cool chamber (ZECC), which could not have been possible without scientific planning, experimentation and validation, preceding product development.

Almost two decades before Prajapati's Mitticool refrigerator, Dr Roy, a scientist with the Indian Agricultural Research Institute (IARI), had developed the ZECC, an innovation aimed at extending the shelf life of fresh fruits, vegetables and flowers at farmyards itself. It did not get much publicity, though it addressed a vital challenge. Was it the timing or the fact that coming from a research institute it did not match the excitement of an innovation from a school dropout? Maybe both. Both these frugal innovations were based on the same theory of the evaporative cooling property of water. Prajapati got the idea from the age-old tradition of using a matka or surahi, which is an earthen pot made of baked porous clay to keep water cool during summer months, and used it to build a cooling cabinet (which brings down the temperature from ambient temperature by 5 to 6 degrees Celsius) to keep household food and beverages cool and fresh for a couple of days. Backed by the NIF, this innovation was picked up for the Innovation Award, 2009 and promoted well in India and abroad. It was widely adopted by local potters of Gujarat and was well received by the BOP users.

Dr Roy, being a horticulture scientist, was concerned about the challenges farmers faced in keeping their harvest of vegetables and fruits fresh for seven–eight days or more to stagger the sale of the produce and get a better price in the market rather than glutting the market with all the produce at the same time.

After years of experimentation, multi-location trials and refinement of the structure, Dr Roy and his team devised a simple double brick-walled chamber in which the space between the two walls is filled with sand and a trickle of water from the top ensures that the sand is kept just moist enough. The chamber is protected by a porous cover and shade, allowing good ventilation. In a dry and hot climate, this can bring down the temperature by 10 degrees Celsius or more, increasing the shelf life of fresh vegetables, fruits and flowers by seven–ten days. This truly frugal innovation was picked up by Dr A.P.J. Abdul Kalam in 1998 for Vision 2020 of Technology Information Forecasting and Assessment Council (TIFAC) as one of the top 50 innovations of the last millennium.[13] However, its diffusion remained limited due to lack of incentives, the nature of the technology and policy support. Since farmers had to construct the ZECC themselves on their land, execution of technology without an expert's supervision posed a major problem. Often, they were unable to construct the chambers according to technical specifications, leaving a wide gap between its expected and achieved performance. But wherever technical support was available, this innovation is in use by the farmers in different parts of India, even after thirty years. This also includes some chambers with modifications to suit the local needs.[14] Therefore, frugal innovations riding on locally available resources, lean design, best practices, efficiency and with the backing of science are more likely to be sustainable.

Gurtej S. Sandhu, an IIT Delhi alumnus, and VP, Micron, is known for his significant contributions in the field of high storage ICs. He was named the seventh most prolific innovator in the US. He says that 'an innovation has to be affordable to be a success'.[15] In other words, irrespective of the economic status of the target users, in today's competitive world, every innovation must be built around the philosophy of frugality.

Science Before Innovation or Innovation Before Science?

While frugal innovation must be encouraged, caution is needed to make sure that these are based on sound technical grounds and are safe to use. Safety assessment is required in traditional innovation to ensure that the final product will not cause any harm to the users and society at large. As Balkrishna Rao, a proponent of scientific rigour in frugal innovation, says, 'After the successful vetting of a new idea for frugal innovation, accurate scientific models need to be adopted for design, engineering, and materials usage, sustainable manufacturing, and testing the product using all probable scenarios, including low probability events.'[16] Whether an innovation is the result of a systematic plan of work from ideation to product launch or simply a spontaneous effort to deal with a problem faced by an individual or a community, including the 'sink to source' approach of grassroot innovations[17] and the importance of their scientific validation cannot be ignored.

Though most frugal innovations result from an understated scientific inclusion, in some, the innovation could happen only as a result of planned scientific research, designing and validation. Development of multimodal stable microscopic platforms for biomedical imaging by the IIT Delhi team is one such example.

Stable Microscopic Platforms for Biomedical Imaging—A Science-centric Innovation

The strong optics group at the department of physics in IIT Delhi is a playground for interesting and innovative work. The user-inspired research of faculty scientists, like Prof. D.S. Mehta, at the bio-photonics laboratory attempts to address either the technical gaps in the marketplace or develop frugal innovations that can have an impact. For instance, this group has designed and developed highly stable, common path digital holographic microscopy (DHM) and speckle-free Linnik-based quantitative phase microscopy (QPM) systems. These interferometry-based advanced microscopes are utilized for the measurement of various biophysical parameters, such as 3D-phase map, 3D height map, refractive index, haemoglobin concentration, dry mass and nano-metric cell membrane fluctuations of various biological cells and tissues without using any exogenous contrast or fluorescent dye. The developed holographic microscope can be used for early-stage disease detection, quantification of sperm cells, oxidative stress condition in macrophages, and health monitoring of red blood cells. Prof. Mehta could establish strong collaborative research in this area with Prof. B.S. Ahluwalia's group at the University of Tromsø in Norway, where the same system was developed and is being utilized for various biological studies. A joint Indian patent[18] on the development of highly stable quantitative phase microscope for biomedical imaging was filed, and there are plans for the system to be commercialized. One of the most important requirements for a DHM or QPM system is high temporal stability (unaffected by external vibrations) for early-stage detection (by accurate measurement of membrane fluctuations of the biological cells) of various diseases in humans. Compared to the conventional DHM or QPM based on Mach-Zehnder or Linnik-type interferometers, the present system offers an energy-efficient and a cost-effective solution. The potential applications of the present innovation can be in the field of profilometry and quantitative phase imaging of various industrial and biological objects.

The Story of Basmati Rice Improvement

In the field of agriculture, frugal innovation has always been the guiding principle of all scientific research. There is a misconception about lack of scientific rigour in innovation made in agriculture. The progress that India has made in agricultural research in the last fifty to sixty years could not have been possible without a strong framework of scientific research, complimented by dedicated team work. The story of Pusa Basmati 1121 (PB 1121) will dispel that myth. It is an example of how thirty-five to forty years of planned research resulted in combining the exquisite grain quality of basmati rice with high-yield, short duration and short plant height.

Basmati rice, the pride of the Indian subcontinent, is a group of long grained, slender and aromatic rice varieties that get its unique taste and texture from the geographic conditions prevailing in the plains of the northern Himalayan foothills. A combination of long grain, appealing aroma, fluffy texture, excellent taste, easy digestibility and longer shelf life makes it extremely popular among consumers (Singh et al, 1988). In India, basmati rice is primarily grown in the Indo-Gangetic region of the north-western region comprising Punjab, Haryana, Himachal Pradesh, Uttarakhand, and Jammu and Kathua districts of Jammu and Kashmir, and twenty-seven districts of western Uttar Pradesh, which have been earmarked as the geographical indication (GI) for basmati rice. The GI status was conferred to basmati rice in 2016.[19]

Traditional basmati rice varieties are very tall, prone to lodging at maturity (due to long and weak stems), photoperiod sensitive and temperature sensitive, which means they have a very narrow geographic adaptability for cultivation and poor yield. Therefore, their production was limited earlier and it could not become a commercial success in spite of global preference and market demand. Seizing the opportunity, the basmati rice breeding programme was planned at New Delhi-based IARI, popularly known as Pusa Institute.[20] Pioneering research work was carried out at the Indian Council of Agricultural Research (ICAR)-IARI, starting from the late 1960s on standardization of protocols for estimating various basmati quality parameters, analysing inheritance patterns, selecting the right combinations for breeding and finally developing the world's first photoperiod insensitive, medium height, and hence lodging resistant and high-yielding basmati variety Pusa Basmati 1 in 1989. Starting with the hybridization of Basmati 370, a traditional low-yielding, tall, long-grained basmati variety from India, and Taichung Native-1 (TN1), a high-yielding, medium height, bold-grained and sturdy rice variety from Taiwan, it took a series of successive crossings and selection of the desired traits, involving nine to ten genotypes to finally obtain Pusa Basmati 1 (Singh, et al, 2018). [21]

It was a stupendous success. Encouraged, the team continued breeding for improvement of basmati varieties with better aroma, shorter growing periods, higher yield, better grain quality and also resistance against major diseases. It applied conventional breeding strategies, along with various molecular tools. This resulted in the development of an extraordinary variety PB 1121 in 2003, which has the world's longest cooked grain length. This was a major breakthrough. However, the saga of basmati rice did not end there. Breeders continued to work to develop high-yielding and aromatic rice varieties which give more income to farmers and 'better taste for less price' to consumers. So, came Pusa Basmati 1509, a variety that matures almost one month earlier than PB 1121, being early to mature, farmers save on at least one round of irrigation and get better price in the market; and PB 1718, a variety that is bred to be tolerant to one of the serious diseases of rice, bacterial blast, by using molecular tools, thus protecting farmers from huge losses at no extra cost![22] With each new variety, productivity increased and the cost of production reduced significantly. All this was achieved at a very modest budget of about Rs 2 crore over a span of ten years,[23] which by any standard qualifies to be called frugal. The product is so outstanding that it was lapped up by dozens of private seed companies in no time and through them, it spread to premium brands of rice sellers. Today, even after sixteen to seventeen years, PB 1121 is the ruling variety when it comes to basmati brands such as Maharani, India Gate, Daawat, Patanjali, Tilda and many more. PB 1121 is an example of innovative application of pure scientific tools in the development of a frugal technology and it is the most popular basmati rice in the international market today. Rice worth more than Rs 20,000 crore is exported every year.[24]

Whether prompted by scarcity of resources or not, the proponents of frugal innovation look at the bigger picture and attempt to create cutting-edge technology without necessarily investing in the use of high-end technology and cutting down on any non-essential frills. Academia and R & D organizations having multidisciplinary expertise and necessary infrastructure are the best places to provide technological backstopping and upgradation opportunities, along with financial and regulatory support to innovations. Strong research-oriented institutions like the Indian Institute of Science (IISc), Bengaluru, and IITs—Bombay, Madras and Delhi, have some of the most supportive technology business incubation (TBI) centres in India. These centres provide an entrepreneurial environment to budding innovators, while scientific expertise at parent institutes helps them convert their ideas into useful innovations. In all this, science is foremost, which needs to be applied innovatively to develop, inter alia, frugal technology solutions. In technology development, it is almost impossible to delineate the boundaries of science, innovation and technology.

Realization of this fact and its internalization has brought several changes in the way we practice science now. The scientific institutions of higher learning have shed their image of being elitist or exclusive, and now actively participate in guiding young minds even from schools. Many institutions have opened doors to students from secondary and senior secondary classes to try their hands at science projects. They have also organized science days and open days, allowing schoolchildren to interact with the faculty scientists and learn more about the whys and hows of various technologies. In many cases, exposure to such S&T institutes like ISRO, IITs, IARI, All India Institute of Medical Sciences (AIIMS), etc., have encouraged these young minds. Akash Manoj's frugal innovation is a case in point. Manoj from Hosur, Tamil Nadu, was in class VIII when he lost his grandfather to a silent heart attack. It left an impact on his young and inquisitive mind and made him wonder why there was no simple device to detect such a risk. Consulting books and experts, including scientists of IISc, helped him understand the problem and devise a small silicon patch which can be stuck on the back of a person's ear or wrist to monitor the occurrence of a silent heart attack. The silicon patch uses an electrical impulse which is positively charged and attracts molecules that are negatively charged, such as fatty acid binding proteins (FABPs) 3, towards it. A person at risk of a heart attack is expected to have a high level of FABPs 3, and therefore he/she can be provided timely medical attention. Due to his scientific merit, Manoj was given research facilities at IIT Madras and AIIMS to develop his innovation, for which he received National Child Award for Exceptional Achievement, 2017.[25]

It becomes evident from various examples of frugal innovation discussed above that whether it is driven by the passion of a young researcher or an institutional goal to develop a better technology at affordable cost, frugal science was neither before nor after innovation. It was embedded in each of these as the guiding philosophy.

The days of meriting only theoretical knowledge as the mark of excellence, particularly in the fields of applied sciences, be it engineering, biopharma, medicine, ICT, computers or agricultural sciences, are gone. Students are encouraged to come up with 'out-of-the-box' solutions to conventional textbook problems, and due credit is given to problem-based projects. Formal partnership of scientific institutions with respective industries in public-private partnership (PPP) mode is showing encouraging results too. Many young graduates from the IITs, IISc and other research-based institutions, and conventional, professional and private universities, and top-ranking B-schools are opting for their own start-up ventures rather than securing high-ranking jobs. There are several factors responsible for this change, but science, or rather its unique applications in finding commercially viable and sustainable solutions with state-of-the-art products and services, has been the biggest game changer. This, coupled with a global preference for 'no-frills' and affordable frugal solutions in every sector, gave rise to a movement of a sort in the present millennium, which caught the attention of the people at large in the last decade after being spotlighted in *The Economist* in 2010.[26]

The importance of frugal innovation, as a powerful option for sustainable growth, does not need to be emphasized. Though the notion of frugal innovation initially originated from grassroots' innovations mostly in India and China[27] and other developing economies, it soon encompassed a wide range of innovations, from rural electrification to medical diagnostics and particle physics, with some based on deep scientific interventions and others redesigning an existing technology to make it more affordable. These innovations are often called advanced frugal innovations, and are low-cost sophisticated products made from minimal resource use.[28]

Many frugal innovations may be built on existing solutions and inventions, not by reinventing, but by improvement, suitable modification and innovative applications. Since these offer significant improvement in performance at a substantial cost reduction, these are able to create a valuable niche in the market. So, though the old advice of not wasting resources on 'reinventing the wheel' still remains relevant, some of the structural modifications of wheels have brought in significant functional improvement at nominal cost. It has been redesigned, modified, upgraded and restructured through the ages, using a variety of materials to serve numerous divergent purposes. When a farmer redesigns his bullock cart wheel by fitting it with tyres commonly used in a three-wheeler automotive carriage to improve the speed and efficiency, it is certainly an act of grassroot innovation. The redesigning of the bullock cart by a non-government organization (NGO), the Centre for Action Research and Technology for Man, Animal and Nature (CARTMAN), in Karnataka, by providing steel wheels with rubber tyres and making some structural adjustments, improved the efficiency manifold at only about 30 per cent additional cost, is an example of frugal innovation with scientific inputs in it. For rural India, with nearly 1.4 crore bullock carts as the primary mode of transport[29] for farm produce as well as other goods and personnel, such frugal innovations make a difference.

Frugal Innovation with a Vision

Besides immediate reasons like facing personal challenges or having limited resources, there are innovators who look at the bigger picture and make it the mission of their life to find solutions for the challenges affecting people and society at large. They make use of their knowledge and apply it to find innovative solutions that are affordable without being a 'will-do' technology.

Amit Bhatnagar is an atypical techie who started his journey with a strong desire to solve the country's problems. This was evident early on when one of us found him to be impatient on his mission but helped curate his case for support under India's landmark innovation support scheme of yesteryears, the Techno-entrepreneurship Promotion Programme (TePP) of the Department of Scientific and Industrial Research, (DSIR). Bhatnagar is an IITian who did his master's in the US and worked with Universal Studios as a business consultant. He quit that job to embark on a journey he believes is of nation-building.

Bhatnagar's biggest strength perhaps lies in the fact that he was not looking for ideas but for problems that were enervating our society, and he was ready to give several years of his life to solve some of the major problems. So, on his return to India, he spent nearly four months to understand the major challenges affecting the Indian population at large. The problem that caught his attention was people being deprived of quality diagnostic services, which resulted in severe health issues, as people did not realize the reasons behind their medical problems, and thus did not take adequate and timely remedial measures. He could relate to this problem as well since he lost his father to jaundice because it wasn't detected in time. Bhatnagar realized that if they as a middle-class family faced such a sorry situation, the people in rural settings with lower incomes must be worse off. Thus started a journey of innovation for Bhatnagar, who was driven not by esoteric ideas but the real needs of society.

His journey led to a pioneering innovation: 'Lab in a Box'—a laboratory that is portable yet rugged. This diagnostic lab needs no air-conditioning and works on 40-watt power, which is one-twentieth the power required by a conventional laboratory. Interestingly, this portable and foldable lab, capable of performing thirty-six tests, can be monitored for quality and usage even from a distance of 500 km! Bhatnagar's start-up, Accuster Lab, secured the privilege of being launched by then President of India in May 2013. The usefulness of the product can be gauged from its long list of clients—the Indian Army, Border Roads Organisation (BRO), Reliance Foundation, HelpAge, Hope Foundation Vadodara, etc. It works even in remote and difficult-to-reach hilly areas.

The innovation led to another revolutionary variant—Labike, which is capable of performing electrocardiograms (ECGs), has a tent structure and foldable chairs, can calculate the body mass index (BMI), has a capacity for cold storage, biomedical waste management, etc., all integrated into one platform. Most appropriately, the Labike was launched in Manipur in February 2019.

Bhatnagar's journey reflects the true potential of educated and trained Indian youth whose innovative work can address the many challenges the country faces.

Levels of Innovation

Larry Myler (2014) has classified innovators into four distinct levels, in which level-four innovations are rarely frugal but the most path-breaking.[30]

- Level-one innovators: They are problem-solvers. They innovate to resolve their challenges using a 'can-do' attitude, are practical, and can fix most day-to-day problems by applying the 'jugaad' kind of innovation.
- Level-two innovators: They are problem-preventers and more experienced than level-one innovators. They can assess the depth of challenges more realistically, even though their technological capabilities may be limited as

in level one. Level-two innovators look to manage and mitigate the market and operational risks first.

- Level-three innovators: They are those who are dissatisfied with the present state of things, aware that the same can be improved, have knowledge and want to improve by bringing an innovative change. For them, solution is not just found in one shot but is a continuous process of improvement.
- Level-four innovators: They have vision and are fully engaged, entrepreneurial and come up with strategic solutions or introduce a breakthrough that provides a cutting-edge solution.

As far as frugal innovations are concerned, all four levels are relevant in providing solutions under specific situations. Innovators like Bajya are examples of level one, and Mansukhbhai and Prajapati level two, respectively; Dr Roy may be classified as a level-three innovator, whereas Dr Singh of IARI, Prof. Mehta of IIT Delhi and Amit Bhatnagar are typically level-four innovators.

Level-one innovators depend more on jugaad and on locally available resources and support. These innovations may have limited scope for diffusion but are extremely useful under a given set of conditions. Innovators at level two would be expected to use their own knowledge and skill but also benefit from collaboration in attaining incremental success in their innovation like Prajapati. Such innovations have a better chance of adoption with support from organizations like CARTMAN or Honey Bee Network. At level three, the innovation is a result of systematic study undertaken by R & D teams of public or private organizations, targeted for the users with limited means, as in the case of ZECC. For level-four innovations, sound scientific interventions are integral at all stages of development, and users can be across the economic boundaries.

Nurturing Frugal Innovations

The uniqueness of frugal innovation lies in the fact that there are no minimum fixed levels of technology barriers to be crossed, scope of utility to be defined or number of users benefited to recognize the worth of such innovations. Almost anyone can be an innovator, from school dropouts to fresh graduates or qualified S&T professionals. The innovation, thus made, can be extremely useful and crucial for an individual or a community's need without much financial value, or it may be at the level of a disruptive technology promising a market breakthrough. The world today values innovation more than the conventional approach as it offers quicker solutions and greater flexibility in addressing the changing needs of the consumers and hence a bigger market.

However, it goes without saying that the path of innovation and its diffusion among intended users can be widened and made easier with necessary support at various stages of development. Be it a level-one or level-four innovation, initial funding to experiment, technological backup to find the best option; refinement of a marketable innovation and scale up, and support in drawing up a business plan and marketing are needed by all. Understandably, the needs will be of different kinds, depending on the scope and complexity of the innovation, level of the users' receptivity and limitations of the innovator. In the following passages, we will examine the needs of nurturing different types of frugal innovations, options and limitations.

It is well known that problems in resource-constraint situations almost always offer opportunities for frugal innovations. However, while most level-one innovators work out of necessity to solve their own problems, like Prajapati or Bajya, there are others, like rural mechanics, foundry owners and artisans from Haryana and Punjab or Gujarat who see their problems as business opportunities and innovate to provide affordable solutions using alternative technology and or locally available resources. There are even instances where these solution providers become knowledge partners in rural areas and co-create technologies needed on the ground, such as rice transplanters in Odisha. Considering that access to technology is inversely proportional to the distance of a village from the nearest urban locality, a city or a town, just like other energy-based household amenities,[31] these rural techno-entrepreneurs devise simple innovations that can be accessed and used in the village with limited resources and poor infrastructure with respect to roads, electricity, water source, etc.

Support for Frugal Innovations at the Grassroots

In contrast to the routine and repetitive activities in the manufacturing, construction and service sectors, farming is based on experiential ingenuity, which needs situation-specific solutions and calls for spontaneous decisions. Farmers, therefore, often show high levels of innovativeness. Considering the significant scarcity of resources they work with, it is only natural that almost all their innovations qualify as frugal, though only a small percentage of these are able to receive the necessary technological/funding support and reach a marketable stage of development.

Ranjit Mirig, a small-time farmer from Sambalpur, Odisha, was faced with the persistent problem of finding affordable farm labour in time. In this predominantly paddy-growing region, shortage of labour was more acute at the time of transplanting and harvesting. Transplanting of paddy seedlings is a time-consuming and back-breaking manual task. Machine transplanters—both mechanical and self-propelled—are available in the market but priced at about Rs 20,000 and Rs 2–3 lakh, respectively; these are neither affordable nor efficient. Desperate to find a solution, Ranjit developed a prototype of a manual paddy transplanter in 1986 but it was far from perfect. At this stage, Ranjit received support from NIF, which proved immensely valuable. NIF not only helped him improve and test his machine but also connected him to IIT Kharagpur for refinement of the machinery. As a result of this, he built a five-row transplanter that covered an area of 1200 square metre per hour at only about 15 per cent of the time needed for manual operations and twice as fast as self-propelled transplanters. A successful model was ready for upscaling by 2008. With help from Honey Bee Network and SRISTI,[32] Ranjit has filed a patent application and this frugal innovation is ready to go commercial. The level of farmers' innovativeness can be judged from the fact that a recent study by SRISTI showed that out of all patent applications filed for agricultural machinery and farm implements, 37 per cent came from farmers, 36 per cent from manufacturing companies, and 27 per cent from ICAR, CSIR and other similar organizations.[33]

A large number of such grassroots innovations are being developed on the principle of frugality in every part of the world. But most often, these remain confined to limited users and never become popular. In some cases, national, international or intergovernmental agencies like NIF, WHO, United Nations International Children's Emergency Fund (UNICEF), World Bank, etc., pick some of these for upscaling and spreading them to users in other parts of the world. The NIF was established in 2000 by the Department of Science and Technology (DST) with the objective of hand-holding innovations at the grassroots level. It has since helped several of its annual awardees secure an intellectual property right (IPR), improve the design and commercialize the same. All such devices are essentially the result of frugal innovations, borne out of necessity, where the innovators have addressed their problem ingeniously, be it Meher Hussain and Mushtaq Ahmad's 'low-cost windmill for water pumping', which operates even at very low wind speed of 8–10 km per hour and costs only Rs 80,000, or the cyclo-cleaner designed by four class VIII students of Delhi, which can improve the cleanliness drive drastically. The NIF selects most promising innovations for National Grassroot Innovation Awards, presented annually to encourage frugal innovations at the grassroots level.[34]

Ennovent on the other hand, provides institutional support to grassroot innovations. A business innovations catalyst based on the philosophy that sustainable solutions for low-income markets can create long-term business value, it partners with the private, public and other sectors to take novel business ideas to markets in developing countries. Since 2008, Ennovent has accelerated over 250 innovations in thirty countries through fifty projects.[35]

It is encouraging to note that during the 107th Indian Science Congress in 2020 held at University of Agricultural Sciences, Bengaluru, Dr T. Mohapatra, director-general, ICAR, announced setting up of a farmers' innovation fund, which will be used to scientifically validate all farm innovations and provide necessary support to refine, upgrade and document innovations of farmers to ensure that science-based farm innovations are adopted rapidly.

Role of S&T Institutions in Nurturing Frugal Innovations

Innovations thrive on deep scientific foundations and by receiving strong insights from multiple perspectives. For instance, a multidisciplinary team comprising doctors, chemical engineers, chemists, biotechnologists, biologists, biochemists, pharmacologists, pharmaceutical technologists and herbal specialists is likely to be more successful in developing an effective and affordable drug in a shorter time than a team of only world-class chemists or doctors. The purpose to innovate may keep changing, but if flexibility and preparedness to fail is interwoven into the fabric of innovation, a multidisciplinary collaboration rather than a single line of research can better tackle development.

For addressing problems of greater complexity, which require high levels of scientific interventions and aim at developing affordable solutions for the larger benefit of the people, international multi-institutional collaborations leveraging mutual expertise and strengths constitute a very effective option. Rotavac, the rotavirus vaccine, an indigenously developed safe and affordable vaccine effective in preventing diarrhoea and other associated health conditions, is one such example. Rotavirus is one of the leading causes of severe diarrhoea and death among children less than five years of age. In India alone, according to an article in *The Hindu*, around 78,000 children die from rotavirus diarrhoea annually, while nearly 9 lakh children require hospitalization.

Rotavac was the result of a multi-country and multi-institutional collaboration of scientists for over two decades. This social innovation brought together Government of India's DBT; the Indian Council of Medical Research (ICMR); IISc; AIIMS; the Translational Health Sciences and Technology Institute (THSTI); the Society for Applied Studies (SAS); Christian Medical College (CMC) Vellore; King Edward Memorial Hospital (KEM) in Pune; and Stanford University School of Medicine; National Institutes of Health (NIH) in the US; Centers for Disease Control and Prevention (CDC) in the US; Johns Hopkins University, US; and PATH, (formerly known as Program for Appropriate Technology in Health), an international non-profit global health organization. Partnership with Bharat Biotech International Limited was vital in commercialization of the Rotavac vaccine, which was marketed in unit sizes of 0.5 ml for Rs 500, thus making it affordable in developing economies, where it is most needed. The vaccine was launched under the National Immunisation Programme in March 2016 in four states and has now been introduced in every state of India.[36] The scientific competence in vaccine production put India in the lead at the time of the Covid-19 pandemic, which not only manufactured enough vaccines for itself, but also made them available to other countries, mostly in the developing economies.

Supporting Frugal Innovations

As far as institutional policies to promote frugal innovations are concerned, more concerted efforts have been made by research institutions both in the public and private sector in developed countries than in the developing economies. Besides problem solving, affordability and market demand, there are some other equally strong motivations like societal well-being and equity for advocating frugal innovations. Institutions like the Tata Centre of Innovation and Design, MIT, have a mission to address the challenges of resource-constrained communities, with initial focus on India. MIT faculty, graduate students, and researchers work with on-the-ground collaborators to learn about and identify opportunities to develop appropriate and sustainable solutions, which in principle encourage frugal innovation.

　　Development of a technology goes through a number of stages, from conception to being market ready, as defined by NASA (Technology Readiness Levels or TRL 1–9), to assess the maturity of its technologies to be used in different missions, which are now used in common parlance and applied to define almost all types of scientific technologies and innovations. Scientific backstopping is most needed between establishing the PoC (TRL-2) and developing a fully functional prototype ready for testing (TRL-6). Working in close association with scientific teams in research institutions is the backbone of science-led innovation and technology development. Support from scientific institutions is as critical as the backing of a funding agency till the product takes off commercially. Contrary to the general belief that the scientific community, particularly in public research institutions in India, is not forthright in extending support to the innovators unless it is in their own interest, the experiences of a number of innovators are quite encouraging.

Innovation and Technology Incubation

In most frugal innovations, strong scientific collaborations are required to upscale and create a successful product out of it. As rightly pointed out by Balkrishna Rao, the weaknesses of frugal innovations can be substantially minimized by applying suitable engineering design tools rooted in scientific principles.[37]

The universities and institutions of higher learning, including the Institutions of National Importance (INI), are supportive of independent innovators, when an idea or a concept exhibits potential for a viable technology. These institutions, as well as other public and private universities and S&T institutions having strong S&T foundations and adequate infrastructure. Adequate infrastructure may offer their resources to enable research translation and innovation. Successful technology transfer is beneficial to both the innovators and institutions. While the former receives motivation and incentives, the latter adds to its reputation and has the opportunity to mobilize resources from royalties and other spin-off benefits. The virtuous cycle of innovation and their gainful impact on society are becoming a trend, though it may not be widely known. There are hundreds of amazing innovations from around the world, which are products of collaborations with academic institutions. Some of these, like Internet, have changed the world like never before while others have saved lives, but all have had a significant impact on our lives.

Indian institutions may not yet have an equivalent example of what a technology like the Google search engine or the recombinant DNA technology did for Stanford University, or fibromyalgia drug Pregabalin did for Northwestern University. However, many frugal innovations incubated by Indian S&T institutions are fairly successful and show great promise of market diffusion not just within the country, but also a likelihood of gaining traction in other developing countries.

Funding by the programmes supported by various government departments and to some extent by corporates like Tata Trust, Reliance Foundation, Infosys, GE, Unilever, Pfizer and many more, have played a crucial role in supporting the innovation and start-up ecosystem.

Barring a few, the various institutional supports that are available in the Indian scenario are not exclusively designed for frugal innovations but for technological innovations/ entrepreneurship in general. Compared to the situation nearly two decades ago, the country has witnessed a steady augmentation of the support system towards innovation and incubation activities. But in the last five years, there has been a huge surge in innovation and entrepreneurship activities as also in the institutional and policy support towards the start-up ecosystem. Realizing the opportunity of addressing a plethora of socio-economic challenges, many innovators are looking at creating an impact through their work.

Given the development status of a very large part of rural India, its specific needs and growing aspirations further push the search for frugal solutions.

Institutional Interventions

Aside from various direct or enabling policy initiatives (see chapter 7 for details), several other institutional interventions of the last twenty-five years or so that have contributed to strengthening the innovation paradigm in the country. This is not to forget or downplay the foundational work done earlier, but only to emphasize the importance accorded to innovations and start-ups in our current development journey. First, there are several government bodies, including ministries that have come up with specific programmes to support innovation and entrepreneurship activities either to enable innovation infrastructure or provide any grant or investment support to start-ups. Some of the important government programmes and support organizations that have had a veritable impact in helping the ecosystem include:

- **Ministry of Science and Technology:** This, by far, has had the biggest impact on furthering the cause of technical innovations in the country largely from and by academia and research establishments. Its various departments and other organizations that are at the forefront, merit mention:
- **DST:** This department pioneered the establishment of the Science & Technology Entrepreneurs Parks (STEP) during mid-1980 and from the 1990s, and the TBIs under the aegis of its National Science & Technology Entrepreneurship Development Board (NSTEDB). The prominent ones like Tiruchirappalli Regional Engineering College Science and Technology

Entrepreneurs Park (TREC-STEP), Trichy; Science and Technology Entrepreneurs Park IIT Kharagpur; S&T Park at the University of Pune; Society for Innovation and Entrepreneurship (SINE) at IIT Bombay; Venture Centre at NCL, Pune, etc., have been able to support hundreds of innovation projects by start-ups. These have included several social impact entrepreneurs.

- **DSIR:** This department has initiated interesting programmes related to indigenous technology promotion, development, utilization and transfer. It has the sole distinction of having initiated the pioneering support programme—TePP. Under this intervention, money was given directly to innovators and start-ups for their innovative work, but the control, monitoring and overseeing was entrusted to reputed incubator partners in the country. This scheme can truly be regarded as the forerunner of several support schemes in the country for innovators and start-ups. Quite a few TePP-supported grassroot start-ups introduced frugal innovations.

- **DBT:** This department has created excellent scientific infrastructure in its domain, established the Biotechnology Industry Research Assistance Council (BIRAC), which has now come to symbolize what a fine innovation supporting institution ought to do. Its suite of support schemes meeting nearly everybody's requirements attests to its relevance in the ecosystem. Besides operating workhorse schemes like the Biotechnology Ignition Grant (BIG) Scheme, Small Business Innovation Research Initiative (SBIRI), etc., true to its dynamic character, BIRAC also comes out with thematic support programmes depending on the market need. One of its schemes, Social Innovation Programme for Products Affordable & Relevant to Societal Health (SPARSH), is dedicated to creating innovations that can have a direct impact on public health. BIRAC derives its strength by leveraging the reach and capacity of its partners.

- **Ministry of Electronics and Information Technology:** This ministry has created a Technology Incubation and Development of Entrepreneurs (TIDE) programme to support start-ups through TBIs. Its programmes leverage the strength of established innovation and incubation partners. Their support is spread over a broad area of electronics and IT—product or application development, and the support extends to distribution.

- **Atal Innovation Mission (AIM):** This is a flagship foray of the Government of India to promote a culture of innovation and entrepreneurship in the country. It has developed relevant programmes and policies in this regard. A few noteworthy initiatives by AIM include Atal Tinkering Labs (over 5000 established till date), which are designed to create a problem-solving mindset across schools in India; Atal Incubation Centres to foster world-class start-up work in thematic areas; Atal Community Innovation Centres to stimulate community centric innovation and ideas in the unserved or underserved regions of the country including tier-2 and tier-3 cities, etc. Besides this, AIM enables involvement in various sector challenges towards improvement of research capabilities in the start-up and micro, small and medium enterprises (MSME) under Atmanirbhar Bharat Abhiyaan.

In addition to these ongoing programmes, several technology-related departments in different ministries, have been supporting sustainable innovations by funding programmes to varied extents, helping industries produce quality and affordable products or enable solutions for socio-economically marginalized sections through NGOs or self-help groups (SHGs).

In a sphere where innovation development and technology dissemination are enabled, two kinds of institutional set ups deserve special mention: technology transfer organizations (TTOs) and TBIs. TTOs create the necessary conditions for knowledge transfer from technology developers to users, which can be various organizations or even start-ups. On the other hand, TBIs generally build capacity for techno-commercial development through the start-up route.

The country has seen the establishment of over 200 technology business incubators and a few dozen business incubators and accelerators. Most of the active TBIs are associated with academia, a beneficial proposition since incubators are able to leverage the infrastructural and intellectual resources of the university/institute. The prominent incubators have been front ending several institutional support programmes for innovators and entrepreneurs. Quite a few incubator-based innovators/start-ups work on solutions with the objective of social impact—necessary but affordable products or services as highlighted elsewhere in the book.

Examples of prominent incubators in the country include SINE at IIT Bombay, Foundation for Innovation and Technology Transfer or FITT at IIT Delhi, Venture Centre at NCL Pune, Centre for Innovation, Incubation and Entrepreneurship or CIIE at Indian Institute of Management (IIM) Ahmedabad, Incubation Cell at IIT Madras, VITTBI at Vellore Institute of Technology, etc. There are also ecosystems enabled by state governments and private bodies that facilitate high-technology, sustainable and scalable social-impact enterprises like T-Hub (Hyderabad), Startup Village (Kerala Start-up Mission), Sandbox (Deshpande Foundation India), etc. Several co-working spaces and accelerators in the private sector like 91Springboard, NASSCOM 10000 Startups, Innov8, etc., are enabling the growth of business enterprises, but they need to do more to help the social impact space grow, particularly frugal innovations. Of these, Villgro[38] is one that supports social enterprises that tangibly impact the lives of the poor in the country.

Strengthening this organized entrepreneurial ecosystem has been the role of risk capital. Though only a small portion of that capital goes into social enterprises, good business models, even if in the social impact space (e.g., microfinance), have attracted good traction in the investment circles.

A lot of development and innovations are a consequence of transfer of research resultants from knowledge generators to entities who would put them to practice. It's here that the role of intermediaries like TTOs becomes important. Generally, academia and publicly funded organizations create such entities to enable technology transfer for the greater good of the society. India has a few such organizations. National Research Development Corporation (NRDC) is the oldest body created by the government for such purpose. It has several successes to its credit, taking affordable and indigenous solutions created by public sector organizations to the market place, such as disposable blood bags, indelible ink, small horsepower tractors, super-absorbent Hydrogel, etc. Similarly, DBT created Biotech Consortium India Limited (BCIL) to accelerate biotechnology commercialization. However, from academia, the prized position certainly goes to IIT Delhi's FITT—a self-sustaining and comprehensive TTO that expands IIT Delhi's outreach and leverages its vast knowledge base to create value through industry engagement, technology commercialization and start-up incubation. Its flexible formats allow it to build capacities and involve all stakeholders in the development process. For more than twenty-five years, FITT has been the bulwark of best practices in capturing ideas and creating value for the larger societal good. Its name clearly suggests its raison d'être and diffusion of, inter alia, frugal innovations or enabling support towards them. It actively scouts for buyers for its various patented technologies, of which frugal innovations form a good proportion of its portfolio. As an example, FITT has licensed biogas compression technology to many small organizations in the rural sector. It has also licensed the now popular and affordable smartcane for the visually impaired. A smartcane is an improvement over the standard white cane to assist visually impaired persons. A white cane cannot detect overhanging objects like tree branches, signboards, open glass windows, etc. Also, at times, using a white cane can result in scratching of a parked vehicle or bumping into another person. A smartcane, with its ultrasonic sensor and a firm hand grip, can help solve the challenges mentioned above and empower the user with independent and safe mobility. This innovation not only received the national award from the Ministry of Social Justice and Empowerment and Manthan Award in 2015 but also earned wide acceptance from over 20,000 users.[39]

While the examples mentioned above have presented the frugal innovation initiatives taken by various institutions in India, there are noteworthy initiatives being taken by a variety of institutions around the world. The following examples will help to explain how the cause of frugal innovation is being fostered:

- **D-Lab at MIT:** The MIT D-Lab works with people around the world to develop and advance collaborative approaches and practical solutions to global poverty challenges. The centre offers several courses, workshops, capacity-building activities and research through experiential learning, real-world projects and community-led development.[40]

- **Kenya Hub:** The Centre for Frugal Innovation in Africa or CFIA is affiliated to CFIA Netherlands, an organization founded in 2013 through a strategic alliance between three Dutch universities: Delft University of Technology, Leiden University and Erasmus University Rotterdam. CFIA Kenya adopts a multidisciplinary perspective to develop analytical knowledge and thinking on innovation in resource-constrained environments. The core research focuses on four strategic domains: energy, health, water and agri-food.[41]

- **Frugal Innovation Hub at the Santa Clara University's School of Engineering:** This serves as a liaison between engineering students, faculty members and organizations to solve humanitarian problems. The humanitarian projects that the hub has facilitated are in various countries like Ghana, Gambia, Togo, Benin, Nigeria, Egypt, Uganda, Rwanda, Malawi, Tanzania, Nepal, Indonesia, Philippines, Afghanistan, Mexico, Honduras, Haiti, Puerto Rico, Nicaragua and India.[42]

Today's Innovation, Tomorrow's Problem

Nothing gives a permanent solution to the problems in the world. And hence, mankind will keep using ingenuity and intelligence to find new innovations to tackle newer challenges. Take the example of polyethylene (PE) or polythene sheeting, the wonder innovation of the twentieth century. On 27 March 1933, when R.O. Gibson and E.W. Fawcett, two organic chemists working at the Imperial Chemical Industries Research Laboratory, invented a white, waxy substance that could be heated and moulded easily to any shape and thickness, they could have never imagined that this versatile material would revolutionize the world. From being used as an underwater cable coating and a critical insulating material during Second World War to numerous medical, packaging, construction and multipurpose use for a variety of things, PE became an indispensable item in our lives.

Polyethylene is the largest volume polymer produced globally, with about 100 million metric tonnes per year.[43] And in less than 100 years, it has become a monster threatening the very existence of this planet. Though one can argue that the problem of plastic is more a result of its irresponsible and indiscriminatory use than the plastic itself, the fact remains that it is choking the only planet we have. As a result, one of the most active areas of scientific research at present is to find alternative uses of plastics and develop biodegradable, compostable or oxo-degradable substitutes of PE and plastic and discover such microbes that can digest the polythene and its many derivatives.

Therefore, the virtuous cycle of innovation to manage and advance our systems and processes will continue to engage us. Or we may continue to remain befuddled till perhaps the scientific advancement reaches a stage where tools like AI take charge, predicting in advance the various possible hazards, a timescale of their severity and ways and means to manage before a technology turns into a menace. Today, we do not have the luxury of spending a fortune in the name of science, unless we promise a solution for large sections of the society that will be effective, affordable and sustainable. The draft Scientific Social Responsibility (SSR) Policy, 2019, of the Government of India proposes to ensure that every scientific research is aimed at providing the best solutions to meet the challenges, fulfilling needs of the society and granting self-sufficiency in a sustainable manner. Undoubtedly, there is no better alternative to science-led frugal innovations.

Successful development of a frugal innovation depends as much on the innovative approach, as the core scientific competencies of the innovator. Institutions and programmes, like those mentioned above, play an enabling role in the promotion of affordable and sustainable solutions by facilitating science-led frugal innovations. Many of these centres, discussed above, not only provide laboratory facilities and physical infrastructure, but also carry the onerous task of connecting the innovator with mentors and experts from ideation to prototype development, and then enable market entry and scale up with requisite links in business and finance. The above steps are in the right direction, but it is still a little early to see the real impact of the measures taken by the institutions to create and popularize frugal innovations.

In the age of rapid technological obsolesce and short development cycles, many novel applications and co-creation of innovative technologies can be expected. Some of these may enable the development of new fields and disrupt existing platforms or business models, whereas others may remarkably enlarge the scope of application of existing technologies by bringing down the cost, using simpler operating models or using alternative resources; or all of the above. No matter how technologies develop or customer preferences evolve, the future certainly belongs to frugal innovations, particularly the ones with strong scientific endorsement.

3

A View of Successes and Failures

If you're not prepared to be wrong, you'll never come up with anything original.

—Ken Robinson

Frugal innovations are considered most relevant in the context of BOP consumers, wherein the commercial success is attributed to the large volume of business.[1] Therefore, low cost and affordability not only make an innovation useful to a large section of population, but also ensure profitability. It is because of this characteristic that some attach high importance to the commercial success of a frugal innovation, whereas, for others the success may be defined by its societal value or ability to resolve a particular problem. However, there are only limited examples of truly frugal innovations, like Jaipur Foot, which have significant commercial as well as societal impact. Similarly, the role of scientific interventions in the innovation of a frugal technology may not be very evident in all cases but, their importance in making successful innovations, be it a rotavirus vaccine, dengue detection kit, Lab in a Box or a super basmati rice variety, in one shot or in incremental steps, cannot be undermined.

What makes an innovation a success or a failure?

In order to answer this, we need to understand some fundamental matters. First of all, how do we measure the success of an innovation? Is it the commercial viability, the cost of the product, complexity or severity of the problem it solves, number of people using it or the length of time that an innovation remains relevant?

Mashelkar and Pandit (2018) have defined seven key criteria for a transformational innovation and called it ASSURED— an acronym for Affordable, Scalable, Sustainable, Universal, Rapid, Excellent and Distinctive.[2] According to them, environment, energy and employment are the three vital parameters on which the success of technologies are to be measured. But as pointed out above, one might argue that all frugal innovations may not fulfil the criteria of universality or scalability, whereas the parameters of measuring excellence might be different in case of say, a grassroot innovation and a science-backed innovative technology. On the other hand, factors not covered by these criteria may play a bigger role, such as cultural diffusion and social acceptance or simply, the relevance of an innovation at a given period of time. Which was a bigger game changer, the wheel, which has lasted many millennia and is still being innovated upon for newer applications? Or the Internet, the broadest and the fastest technological revolution, which not only erased the barriers of information, distance and time, but also changed human behaviour forever?

And if an innovation is not successful in terms of generating good business, should it be termed unsuccessful? Examining the criteria and factors that determined the success of some S&T innovations in the recent and not so recent past might help us understand this complex phenomenon.

There are many misconceptions surrounding the nature, social acceptability and scientific value of frugal innovation. It is generally believed that frugal innovations are meant to solve problems and challenges faced by people with limited resources. While it is true that affordability at the bottom of the ladder is key to success of a frugal innovation, we also need to acknowledge that every consumer, irrespective of his/her purchasing power, wants value for money. Therefore, a frugal innovation that is not inferior to its more expensive alternatives will be adopted by users across economic strata. In fact, by adopting a frugal approach, a world-class cutting-edge technology can be developed at a fraction of the cost and can directly contribute to sustainability.

When a frugal innovation, based on a sound technology, is successful in fulfilling customer needs in developing countries, it may be expected to be successful in developed countries as well. For example, the portable ECG machine (MAC 400) developed by GE for emerging markets sold well in over 100 countries, including some developed markets like US and China.[3] In this case, good technology, affordability and global acceptability all contributed to its success. Possibly, the fact that it came from GE, a company of high repute and credibility, might also have contributed to its success. Recognizing the market potential for portable ECG monitors, particularly for real-time wireless monitoring in ambulances, healthcare centres in remote villages and difficult terrains, as well as at home in the times of war or natural calamities, this device encouraged many small players and start-ups to innovate cheaper, and at times, better substitutes. An Indian firm, Agatsa Software Pvt. Ltd., introduced credit-card sized ECG monitors SanketLife, which measures ECG in 15 seconds. This twelve-lead small ECG device contains a Bluetooth-based circuit that connects to smartphones or tablets, and results can be sent instantly via email, social media platforms, or SMS for information or medical consultation and is priced at Rs 2500.[4] Utility of such devices increase significantly in epidemic- or pandemic-like situations like the Covid-19 pandemic, which forced millions to remain confined to homes for extended periods.

All cheaper and affordable products or services do not necessarily imply recourse to frugal technology. Developing an affordable version of a product using cheaper/inferior components, or a less efficient technology might prove to be a 'penny wise, pound foolish' substitute. In the case of frugal innovation, there is no compromise on technological parameters and this aspect is of utmost importance. More often than not, frugal technology takes years of continuous research, experimentation, successive improvements over the original innovation, many rejections or partial acceptance of the technology by consumers, multidisciplinary collaborations and partnership between many stakeholders before it reaches an acceptable stage. But once it clears all the hurdles, it not only results in a product that is commercially successful, but also brings in due recognition to the innovator. And above all, it reinforces the argument of application of scientific approach in development of frugal innovations. Often in scientific research, the primary objective is to find a viable solution to a problem and develop a technology with specific criteria. However, after achieving the primary goal, and anticipating its potential use in everyday life, researchers work to bring down the cost of the technology by various substitutions, alterations and optimization to make it a commercial success.

A recent example shows the evolution of an innovation, which was founded on an excellent technology and resulted in a frugal product that made an enormous impact on the lives of the people worldwide. The Royal Swedish Academy of Sciences conferred the Nobel Prize 2019 in Chemistry to the trio of John B. Goodenough (University of Texas, Austin, US), M. Stanley Whittingham (Binghamton University, US) and Akira Yoshino (Asahi Kasei Corporation, Tokyo, Japan) for their extraordinary contributions for over three decades in innovating the versatile rechargeable and powerful lithium ion (Li-ion) batteries. These batteries have changed our lifestyle to such an extent that imagining a world without them is difficult today.[5] More importantly, this vital innovation is affordable even to people at the BOP.

Li-ion batteries have actually made the success of many more innovations possible. These batteries are literally at the heart of mobile phones, laptops, e-vehicles and spacecraft. These batteries are not only lightweight, but also capable of storing significant amounts of energy even from solar and wind power, promising a world free of fossil fuel.

The Journey of a Battery

In the 1970s, Stanley Whittingham discovered a highly energy-rich material, titanium disulphide, used it as a cathode along with metallic Li as anode. The 2-volt battery thus developed had great potential but was not safe for mass use.

John Goodenough predicted that the cathode would have greater application if a metal oxide were used in place of metal sulphide. In the 1980s, he demonstrated that cobalt oxide with intercalated Li-ions could produce as much as 4 volts.

Based on these, Akira Yoshino created the first commercially viable Li-ion batteries in 1985 using cobalt oxide as cathode and petroleum jelly as anode, both having intercalated Li-ions.

After many years of research, development, testing and validation, Li-ion batteries entered the market in 1991. But research continues till today for subsequent improvement in terms of storage life, efficiency, safety, environment protection and cost reduction.

The story of converting a superior technology into a frugal and sustainable one and turning it into a successful business highlighted the role played by academia and industry-led R & D, which perhaps was vital to the commercialization of the first viable product in the market and its subsequent improvements. It also brought forth the hard fact that it takes years of persistent and meticulous research and continuous hard work to perfect a technology before a frugal innovation becomes commercially successful. An innovative approach in scaling up and offering a powerful technology at the right time are important determinants in the success of frugal innovation. The power of innovative technology was evident due to a surge in affordable smartphones with unique and attractive features and powerful hardware, which have flooded the markets in recent times, both in developing and developed worlds. Competing brands attracted consumers at the BOP by providing better camera features, attractive exterior, etc.

Many S&T-based start-ups, both standalone ventures or those mentored by academic institutions or the industry, show high potential to be commercial successes but only a few become successful, whereas others fade out after making an initial buzz. However, many grassroot innovations that solve specific problems, such as motorcycle tractors, mahua flower gatherers, foldable solar panels, cotton-stripping machine, maize-cob sheller, bamboo-stripping machine, areca nut peeler, bicycle-mounted field sprayer and many more, are being made in every part of the world by people who innovate to solve problems crucial to their dire needs. Similarly, engineering artisans from small towns and suburbs of Haryana and Punjab, are famous for their jugaad in duplicating complicated machines (even imported ones) or a component for it at a fraction of the cost of the original. But mostly, such innovations remain confined to limited users and hardly achieve commercial scale or success; nor are they intended to do so. In some cases, national, international or intergovernmental agencies like NIF, WHO, UNICEF, World Bank, etc., or private or public institutions pick frugal innovations that they can replicate and distribute to users in other parts of the country or world. The fact remains that barring a few examples, like Prajapati's Mitticool (see chapter 2 for details) or solar grandmothers (see chapter 7 for details), which spread to offshore markets, these innovations remain confined to local populations. But, by no means can these be considered failures! Success of frugal innovation often depends more on socio-cultural parameters than its technological advantage alone. For instance, there is a big market for frugal innovations in Africa, but it is a challenge to launch products for BOP customers there as only 30 per cent of the population has access to electricity and only 34 per cent of the population is accessible through roads.

Fading Out and Failures

Every technology has a shelf life. On 10 March 1876, when Alexander Graham Bell spoke into the first telephonic device to his associate, 'Mr Watson, come here, I want to see you', history was made. This disruptive innovation changed the whole idea of communication. Though the components and technologies underwent many makeovers and refinements to enhance efficiency, reduce cost and retain original voice quality, no one could have imagined that the traditional telephone, which had over 50 million users within seventy-five years of its introduction, would ever become obsolete! In today's world, the turnaround time for development of new technology and innovative applications has reduced so drastically that it is unthinkable that a technology can remain relevant for more than fifty years. Today, the lifespan of even disruptive technologies can be confined to a decade or less. Many disruptive technologies of the twentieth century, such as televisions, cellular phones with punching keypad, photocopying devices and table computers, have become or are on the brink of becoming obsolete today. And they were all based on sound scientific technologies and were highly innovative when commercialized. Though introduced as technologies for rich people, additional refinements and scaling up made these highly affordable within the span of a few years.

Yet, to keep pace with human imagination, these needed to be replaced by more innovative and superior alternatives. So, every technology has a limited shelf life, and to retain the relevance of technology longer, it needs to be upgraded or redesigned constantly in line with the expectations of the consumers.

In spite of the best ecosystem, an innovative technology might not be successful commercially, and the reasons for that may be diverse. Mashelkar and Pandit (2019) tried to examine the reasons for the failure of innovations based on the criteria of ASSURED. As the path of success is difficult to locate, identifying failed innovations to examine the causes of their failure could give some insight.

The *MIT Technology Review* published a list of the worst innovations from the past two decades, which included e-cigarettes and many others, that had caught people's imagination and were introduced with a bang yet failed to make a lasting impact.[6]

Analysing this, Sean Wise, the founder @SeanWise, concluded that premature scaling, lack of customer discovery, and applying outdated methods were among the major lacunae that led to some of the worst innovations of the twenty-first century.[7] These innovations failed to put customers ahead of investors and were built using twentieth-century entrepreneurial rules. If twentieth-century innovations were shrouded in secrecy till the final product was ready to go to market, twenty-first-century innovations belonged to today's start-ups, which often follow the philosophy of 'done today beats perfect tomorrow',[8] unlike the past, when a product was introduced in the market only after it reached the final stage of development. However, time and again, it is seen that affordability is the key for commercial success. The failure of Segway, a two-wheel personal transporter, is an example of an innovative but not frugal technology. It was introduced with a bang and faded without a whimper. The unaffordability and poor scalability are the primary reasons for the failure.[9]

Closer home, we have Tata Motors, which introduced a technologically superior, fuel-efficient and environment-friendly small family car, Tata Nano, for the price of a two-wheeler, yet it was a commercial disaster in spite of the reputation of Tata Motors. One may cite faulty promotional strategy, wrong timing or even error of judgement in assessing public aspirations and preferences. So there is something beyond the frugality that determines consumers' acceptance of an innovative technology. However, as Harvey Brooks (1994) rightly observed, while science plays a vital role in innovation and technology development, there is reverse contribution of innovations and technologies in furthering scientific advancement.[10] Hence, there is always a lesson learnt even from the so-called failed innovations as these could lead to new scientific advancements.

Institutional Successes

Though India is not yet a global player in developing frugal innovations that could sweep the world market, it has many institutional successes that are recognized internationally, putting India forth as a worthy leader. The frugal approach followed in developing the PSLV and Geosynchronous Satellite Launch Vehicle (GSLV) and other technologies used in Indian space programmes, due to the use of indigenous technology by ISRO, drastically cut down the cost without compromising on scientific standards (see chapter 1 for details). It is one of the finest examples of frugal innovation backed by cutting-edge science.

Similarly, we have many success stories of frugal innovation by an organization in agriculture, a sector where independent India has made the most significant progress, both in product and systems innovation. In these, good science led to good technology development at a fraction of the budget allocated in other countries for agricultural R & D.[11] Whether it is the introduction of short-statured, fertilizer-responsive and high-yielding varieties of major food crops like wheat and rice or use of biotechnology for protection against insects (pests), with concerted efforts of agriculture extension machinery and right policy support, farmers adopted these technologies quickly and willingly. All of this played a role in taking India from a food-deficit to a food-surplus nation.

Besides the success stories of green revolution (transforming the cereal production scenario in India), white revolution (milk production), blue revolution (fish production and aquaculture) and rainbow revolution (fruits and vegetables production), India's agriculture sector maintained a higher contribution to gross domestic products (GDP) than other sectors, mainly due to continuous and all-round innovations and their adoption to meet the changing demands. The ICAR, which has an annual budget of about Rs 8000 crore, in partnership with agricultural universities in the country, has developed many such innovative technologies, each of which gives an annual return of Rs 10,000 crore to Rs 20,000 crore.[12] Some of these, like high-yielding basmati rice and raw cotton are key foreign-exchange earners too. The products became popular globally on the merit of their quality, with hardly any business promotional activities behind their success. So, what could be the possible factors that determine the success of a frugal innovation? Let us make some logical assumptions.

Success of an Innovation

Though it is difficult to draw a roadmap to success when it comes to frugal innovations, some basic presumptions and criteria support their acceptance and help market diffusion.

First and foremost, true to its tag, affordability is the key consideration. Affordability ensures acceptance of a frugal innovation. Therefore, the minimum cost at which the basic need of the customer gets fulfilled should be the starting point for organizations to conceptualize a solution. For example, in India, Aravind Eye Hospital, by offering an innovative treatment, patient management strategy and cutting unnecessary costs, offers cataract surgeries at a fraction of the cost charged by other hospitals. The surgery is also free for poor patients (see chapter 5 for details).

Second, it is important to work on a distinct science-based technology with an understanding of the customers' perspective. Investment in scientific R & D leads to technology upgrades and future capabilities that are transformed into new products, processes and services. Consider the success of Intel. With its massive investment in R & D and never-ending ability to change or improve one or two vital components, it adopted a 'tick-tock' policy, alternating between improvement and innovation, and continued to introduce new and better products, leaving its competitors safely behind.[13]

A product that is originally designed for developed markets may require some redesigning and modifications to be successful in emerging markets. It has to keep the customers' choice and their purchasing power in mind. The same applies if the product goes from a developing to a developed market. The social and cultural norms, ethical values of customers and available infrastructure need to be borne in mind while redesigning solutions. Tata Swach, a non-electric water purifier developed by Tata Chemicals, is a perfect example. This modestly priced purifier, which does not require electricity or running water in the user's home, was successful in class-B towns even though models of better standard were preferred by the upmarket segment in bigger cities.

Timely intervention is the third key issue. Bt cotton hybrids were the first (and till date the only) genetically modified (GM) crop released for commercial cultivation in India in 2002. Since the price difference between Bt and non-Bt seeds was more than fourfold, presence of spurious and substandard Bt cotton seed in the market (69 per cent in 2003–04) became a serious problem. In the absence of any standardized and affordable test, quality assurance was a big challenge. So, the ICAR–Central Institute for Cotton Research, Nagpur, devised the dipstick and ELISA kits in 2007 for testing the purity of Bt cotton seed. This technology was commercialized through non-exclusive licensing and became an instant success. The cheap and simple dipstick test came as a boon for both farmers and seed traders. Neither a laboratory nor a qualified technician was needed to conduct the test. However, the ELISA test requires laboratory facility and technical training, and is a quantitative and more precise test for detection and quantitative testing of Cry1Ac, Cry2Ab and Cry1C proteins.[14] The innovator, Dr Keshav Kranti of ICAR, patented these technologies in India and several other countries and commercialized it through non-exclusive licensing. The technology was officially recommended by the Ministry of Agriculture and Farmers' Welfare for quality check of Bt cotton seed. By 2013, more than 40,000 test kits had sold and the demand was growing.[15]

All this was achieved without any funds for business promotion. This success was attributed solely to timely and effective technology diffusion through non-exclusive licensing. However, since the current policy does not promote cultivation of Bt cotton as before, the life of this technology, in spite of being successful, remained rather short—about fifteen years.

Seizing an opportunity, even in the midst of a calamity, can pave the way at times. During the recent Covid-19 pandemic, two of the most sought-after items in the market were affordable hand sanitizers of good quality, which can be used frequently even by children, and an accurate, affordable and quick diagnostic tool to detect Covid-19 infection. Not only was there a scarcity of hand sanitizers as well as protective face masks within the first two weeks of the pandemic in every country, but the quality too was poor and the cost too prohibitive for BOP users. They generally have larger families and poor sanitary conditions of living, hence are at greater risk.

JK Nanosolutions, a start-up based in Bengaluru, devised a frugal innovation to fight the war against Covid-19 in the form of Nano Corona, a nanotechnology-based hand sanitizer with the goodness of neem and active nano silver in 60 per cent alcoholic solvent. As per WHO guidelines, an effective hand sanitizer is required to contain not less than 60 per cent alcohol. However, presence of neem extracts, a germicidal known since ancient times, and active silver with established antimicrobial properties, make this a better choice to provide holistic protection against a range of other infectious diseases, particularly common in developing economies. Nano Corona is available for Rs 300 per litre, much less than the price of other available brands.[16] In such a case, combination of good technology, frugal cost and timely intervention makes this a potential winner.

The challenge of quickly developing a reliable and affordable diagnostic kit for detecting Covid-19 was addressed by many institutions and firms, a few of which achieved different levels of success. While a promising technology, probe-free detection assay, was developed[17] by Prof. V. Perumal's team at IIT Delhi, a Pune-based molecular diagnostic company Mylab Discovery Solutions Pvt. Ltd. formulated a 'make-in-India' kit that was developed and evaluated in record time as per its managing director Hasmukh Rawal. The kit was approved by the National Institute of Virology (NIV), ICMR and Central

Drugs Standard Control Organization (CDSCO). At that time, Shrikant Patole, articulated that they can manufacture nearly 25,000 units of Mylab PathoDetect Covid 19 Test kits per day on ramped up capacity.[18] The company was able to shorten the test time from the prevalent 6–8 hours to 2.5 hours because the kit could do both screening and confirmation simultaneously. Even though the government had initially fixed a maximum charge of Rs 4500 for such PCR-based tests, the Mylab kit was able to substantially lower the costs. What is significant from a frugal perspective is that this kit could work within the testing infrastructure available with Indian diagnostic labs and did not require any new machinery.

Fourthly, in order to ensure diffusion of the innovation to benefit the maximum number of consumers, the innovation must be rugged, human-centric and make use of resources available in abundance and at low cost. The lifetime cost for continued use of the innovation must be kept very low, which implies that repairability and/or replaceability should be easy and not a burden on the customers. The resources already available with customers can be leveraged for easier diffusion. For example, solutions using mobile phones find easy acceptance owing to the high level of penetration among customers even in remote rural areas and emerging economies. Vodafone's mobile phone-based payment service, M-Pesa, received quick acceptance from customers lacking access to traditional banks in Kenya, but in India, with banking services fairly established across the country, mobile-based and e-banking services gained customers' acceptance relatively late.

Institutional Policies and Business Incubation Labs

Policies and institutional support play an important role in creating a conducive ecosystem and hence, the success of innovations in general and frugal innovation in particular. It is true both for public research institutions as well as R & D facilities established by private industries.

Several examples of frugal innovations discussed in this book were formally incubated or provided informal but vital backstopping by leading S&T institutions of India or other countries. It is important that technology management and incubation units hosted by these institutions, besides encouraging their full-time researchers to think out of the box and try their hands at innovating, also support entrepreneurial skills of those having innovative technology ideas. Institutional policies play an important role in nurturing such possibilities (see chapter 2 for details).

Knowledge and technology are the two essential ingredients of innovation. Many scientific institutions have been the cradle of technological innovations and breakthroughs, where a mechanism of formal association with private innovators and entrepreneurs has been very productive. On the top of the list, of course, are Stanford University and MIT. But when it comes to frugal science and innovations, Stanford University clearly takes the lead. Whether we talk of frugal innovations like Foldscope and Paperfuge or breakthrough innovations such as Google and Yahoo, Stanford University was the birthplace of all these innovations. The success of Stanford University can be attributed to its strong scientific base, ability to identify future needs, focus on global issues, promotion of innovations, interdisciplinary team work and a conscious approach to promote frugal science. They also include the public and private sectors from one or many countries to find the best solutions and quick diffusion. However, we need to acknowledge that even with good institutional support and technology back up, a good frugal technology may fail to catch market interest if the time is not right for introduction of such an innovation.

The other type of innovations are the ones that are devised through partnerships between innovators and experts at research institutions (public or private) or by individual innovators or a team from within the institutions (see Bt cotton seed testing kits above). There are many examples from IITs to MIT. Narayan Ramachandran, chairperson, InKlude Labs, once asked, 'What is it about institutions—government, private and social—that allows them to build innovative, low-cost, scalable ideas?'[19] Perhaps, there is no straight answer as factors determining the scalability are as divergent as the innovations themselves. However, it can certainly be said that promoting technology innovation through start-ups and incubating entrepreneurs within an established R & D organization is a win-win model for both. Commercial viability of potential technologies can be tested through start-ups even before these are developed in their final stages, whereas innovators get technological back up and business promotional support from these organizations. It is for this reason that innovation and incubation centres have become integral parts of R & D organizations.

However, the very nature of frugal innovation has an element of non-predictability in terms of their commercial success, which cannot be overlooked. At the time of establishing Aravind Eye Hospital, its founder Dr G. Venkataswamy did not think about its profitability. He simply wanted to provide high-quality eye care to eliminate preventable blindness in India by means of an innovative model of high-volume patient care, which also made commercial sense. Similarly, the goal of frugal science promoter Dr Bhamla's research (see chapter 2 for details) is to innovate low-cost medical devices and other diagnostic/research laboratory equipment, which can be used by all.[20] An assistant professor in the University of Georgia Tech, US, Dr Bhamla once noticed a student in his class tilting his head at a particular angle and learnt that one of the batteries in his hearing aids was not working. The student had been using a hearing aid costing around $5000. The seed of an idea was sown and from this emerged the concept to innovate a hearing aid for $1. Very recently, Dr Bhamla and Soham Sinha, a PhD student at the same university, were able to design and create a 3-D-printed low-cost hearing aid (LoCHAid) that costs under Rs 75 ($1). By September 2020, the device had already complied with five out of six product recommendations laid by the WHO, with work on for its commercialization.

When innovations are an integral part of institutional programmes, even if the technology fails to make a mark, they can be immensely useful in knowledge advancement. The enormous investments made in research and innovations in genetically modified organisms (GMO) and gene editing might not have received public acceptance in many parts of the world, but the knowledge gained and technological progress made by R & D institutions cannot be termed as failures. These innovations are applicable now as well as promise immense possibilities in the future for introducing biotic and abiotic stress-tolerance genes in plants (GM) and correcting genetic disorders and diseases (gene editing), for which no solution is foreseen in the near future. The innovation, by every account, is successful, however, its application and subsequent commercial success completely depend on a conducive ecosystem, decided largely by government policies and informed public opinions.

Research Spin-offs May Not Always Make It

IIT Delhi, IIT Bombay and IISc, Bengaluru, with their strong emphasis on technology incubation and industry partnerships, have been the forerunners in testing, incubating and commercializing innovative technologies. These and other institutions have established dedicated innovation centres for this purpose. Conscious of realizing the investment made, the need to spread technology and add credence to their reputation, these centres follow a system of thorough scrutiny by the best technical experts, validating and assessing the potential worth of proposed technologies before incubating them. While there have been many success stories, there are also cases that deserve a mention for not making it. Such lack of start-up successes can be due to different reasons even when the innovations were bereft of any obvious limitations.

Advantage Organic Natural Technologies Pvt. Ltd. (AONTPL) is a venture with an eye on tomorrow's needs, mentored and incubated at FITT/IIT Delhi. The company launched organic herbal lingerie, innerwear and yoga wear with skin-fortifying finish, developed in technical collaboration with IIT Delhi faculty. Its products are soothing, durable, comfortable and hygienic, with a healing touch for the most sensitive parts of the human body, hence branded as the 'ultimate luxury wellness garments'. AONTPL also has a range of sportswear processed with a concoction of neem, tulsi, apple cider, vinegar and herbal colours, integrating Ayurveda with modern biotechnology, nanotechnology interventions, and bioengineering natural, organic and herbal ingredients. This green/clean technology with a low carbon footprint has already obtained three patents in India, US and Europe.[21] While lack of committed business linkages hindered market growth in the initial years, it was flash floods that damaged the new factory of this start-up and severely impacted its ability to bounce back, a case where an unforeseen natural calamity and other reasons compounded the vulnerability of a technical start-up that should have otherwise found wide acceptance.

Let us examine another innovative research spin-off that could not become a commercial success in spite of having everything in the making.

Many technologies that we use every day consume a lot more resources and power than they need to, and using them creates serious problems in terms of pollution, depletion of resources, ecological imbalance, health hazards, etc. There are many R & D projects that focus on reducing pollution retrospectively and prospectively. An interesting but unobtrusive case relates to pollution-preventing lithographic inks. Pollution in the industry mainly results from clearing the ink from presses. Waste water from printing operations often contains mineral oils, waste ink, cleaning solvents, acids, alkalis, plate coatings, as well as metals such as silver, iron, chromium, copper and barium.

Printing is inevitable because it is an integral part of our everyday activities. Presses need to be cleaned after every batch for the next batch of printing. A common practice is to use aliphatic and aromatic solvents containing volatile organic compounds (VOCs). The commonly used printing inks also contain VOCs, which are emitted into the atmosphere in quantifiable amounts. For example, a typical medium size coldset web offset plant uses 40,000 kg per year of ink, of which 6000 kg is lost as emissions. The plant uses over 1,00,000 kg per year of cleaning solutions, all of which is lost. A typical newspaper uses over 1,00,000 kg per year of ink and emits over 8000 kg of the volatile solvents. All of these result in air and water pollution.

A research group of IIT Delhi, led by Prof. A.N. Bhaskarwar, which is working, inter alia, in the area of clean processes, developed an ink based on natural products, like castor oil. This green ink is completely free from VOCs and thus environment friendly. The group synthesized new alkyd resins and studied in detail the kinetics of washing of this ink with water at a little higher pH.[22] The new ink does not contain aliphatic solvents, so its use produces virtually no emissions during printing. As a result, the printing itself involves no solvent and the presses can be cleaned by washing with water at a slightly elevated pH. Thus, the new ink completely avoids the solvent emissions commonly associated with printing.

The promise of a commercially viable solution led to a spin-off—EnNatura Technology Ventures (P) Ltd., which was incubated at IIT Delhi. This start-up was initiated by a few graduates of the chemical engineering department of IIT Delhi, The start-up was able to secure innovation grants from DSIR and later venture capital investment. The start-up garnered important initial market traction with marquee customer brands and good press.[23] However, the team faced some technical challenges as it tried to scale-up and could not meet the demand. The team decided to close up shop rather than attract a poor reputation for not being able to supply material of consistent quality and in time. The story of a promising spin-off with a strong Intellectual Property (IP) thus got derailed.

This chapter has primarily considered innovations from the organized sector, representing public S&T organizations, R & D-driven industries, or public–private collaborations. However, a large number of frugal innovations, particularly in the Indian context, may be classified as grassroot innovations borne out of necessities of life and developed with little or no institutional support. As observed by Dr Mashelkar, former director-general, CSIR, and founder and chairperson of NIF, 'technological innovations not only emerge from the laboratories and scientific institutions, but also from the laboratories of life'.[24] Though an entrepreneurship model, applicable to the widely divergent situations of these innovators which could convert such innovations into successful enterprises, is yet to evolve, its potential scope cannot be undermined by us.[25] A 10-horsepower tractor costing about Rs 1.25 lakh, designed by a school dropout, Bhanjibhai Mathuria of Junagarh, which was licensed to M/s Pramal Farmatics for manufacturing and marketing under the brand name Vanraj, reinforces such possibilities.[26]

These examples reiterate the unpredictability of frugal technology, revealing that the need to innovate and develop frugal solutions for fulfilling aspirations or mitigating problems faced by society at large, or people of a particular section, does not follow a defined pattern. Whatever may be the objective to create a frugal innovation, its success is measured, among other factors, by one key determinant, that is how many people benefited from it. A jugaad innovation, which is made to solve a specific problem in a given set of conditions, does not target a large number of users. Hence, it does not plan its diffusion in a systematic manner. Even then, many such solutions are adopted by others after seeing and learning from the experiences of the original innovator. The cycle keeps growing, taking more and more people in its fold.

4

Frugal Innovation for Sustainable Solutions

Never before in history has innovation offered promise of so much
to so many in so short a time.

—Bill Gates

The world today is witnessing a plethora of challenges,
from shortage of energy, food and water to grappling with
multidimensional problems of climate change, social inequity,
unsustainable human behaviour or practices and global
pandemics. Despite the grim scenario, these challenges provide
a large window of unique opportunities for thinking out of the
box and pushing mankind to innovate to survive. While some
of the critical problems call for immediate response, including
technological solutions, others can be addressed through
systemic and policy-based interventions.

There are many inspiring examples of how community-
level difficulties have been addressed both by individuals with

modest levels of formal education as well as professionally qualified scientists. Frequent power breakdowns in the middle of the examination season in a village in Bihar led one enterprising and innovative woman to start a facility to charge inverter batteries during the day using a solar panel.[1] This helped the community and also allowed her to earn some extra money for her family. On the other hand, India's fight against dengue and malaria led to the development of rapid and affordable detection kits by innovative scientists at a fraction of the cost of the imported ones. Such situations prompt us to innovate quickly and find solutions that are intrinsically frugal because to compete with an existing one, a new solution must be appreciably economical, no matter how technologically superior it is.

For problems that are bigger in dimension, like climate change and global warming, intergovernmental efforts are needed to find sustainable solutions. And when there is a calamity, like the outbreak of the novel coronavirus, the scientific community of the world comes together. In a meeting held in February 2020 at the WHO headquarters in Geneva, leading health experts, scientists and donors from around the world discussed possible strategies to contain the disease that possibly originated in China towards the end of 2019. Since treatment options were limited, the focus was on development of vaccines and diagnostics. Several vaccines have been launched by pharmaceutical firms that include AstraZeneca, Bharat Biotech, Pfizer, and Johnson & Johnson. In India, more than a billion people have already been vaccinated through a well-orchestrated programme monitored by the Ministry of Health, Government of India.

In the examples we will now discuss, widespread problems led to the development of innovative solutions or ways to manage these problems effectively. In each of these situations, the scientific knowledge used by innovators was vital in providing sustainable solutions to problems of wider consequences. Since many of these innovations, which address community problems, promise little or no commercial value, awards and recognitions instituted by government organizations, NGOs and commercial organizations played an important role in encouraging more and more professionals to innovate and promote eco-friendly and sustainable solutions in different sectors.

The Pollution Warriors

Controlling air pollution requires long-, medium- and short-term government policies and regulations, huge investments and creation of necessary infrastructure. Alongside, it also needs nationwide awareness to promote the use of cleaner fuels in industrial and domestic applications, and adoption of modified systems of transportation and operations almost in every type of development. A lot of these changes will not be immediate and will require adapting, so people will also need solutions to protect themselves from harmful effects. Some IIT Delhi graduates teamed up with their professors to fight the ubiquitous air pollution. Prateek Sharma led Nanoclean Global with tenacity and doggedness to make this start-up a success.

Sharma's campaign against air pollution was inspired in part by his personal challenges. He often saw his asthmatic mother in his home town of Bikaner in Rajasthan struggling to cope with smoke from cooking and also sandstorms. Her dupatta or other available masks were of little help, besides being uncomfortable.

It was at IIT that Sharma discussed the problem with his receptive and enterprising batchmates. That's where the idea of Nasofilters, which are nasal filters, originated. The most important aspect was to develop the filter media since

Nasofilters cover the nasal orifices and it was necessary to ensure that people were able to breathe easily despite using the product. Nasofilters cover the nostrils and not the entire face, aren't uncomfortable to use and protect the user against air pollution.

These IIT graduates looked for scientific solutions and eventually turned to nanotechnology. They reduced the thread diameter of the fabric to a nano size and wove it into a fabric. This resulted in a fabric with pore sizes in microns, even smaller than particulate matter (PM) 2.5. This fabric only allowed oxygen and pure air to pass through, while restricting air pollution by 95 per cent. After multiple researches, iterations and constant support from faculty mentors at IIT Delhi, they were able to develop the world's first biocompatible nanofibre-based filter media. Combining that filter media with the idea of covering one's nose, Sharma and his team members were finally able to develop Nasofilters. The product provides protection against not only PM2.5, but also bacteria, pollen, etc.

The venture was supported by funding from DST and BIRAC. To turn Nanoclean Global into a successful venture, the venture was backed by FITT, an incubation centre at IIT Delhi, and made some of their professors as stakeholders.

Nasofilters were first launched during January 2018 in the National Capital Region (NCR). Within a month, more than 2 lakh units were sold in Delhi. Post that, Nanoclean Global reached out to other states and eventually other countries. At Rs 10 per piece, Nasofilters costs five times less than global competing products (based on 2017 prices). With more than a million pieces sold and the demand growing, it has captured offshore markets too. Nanoclean Global now supplies its product in several foreign countries.

However, the story did not end here. Nanoclean Global realized people spend much of their time indoors and that indoor air can be five times more polluted than the air outside. To solve this problem, the Nasofilters Pollution Net was developed. It is a filter mesh that can be installed on doors and windows instead of the wire/mosquito mesh. Made using the same nanofibre technology, it is designed to prevent pollution from entering homes and has a life of five years.

The Nanoclean Global team also wanted to tackle the problem of indoor pollution due to cooking in the kitchen. Although there are air purifiers to solve that problem, they wanted to provide a more affordable and convenient solution. Frugality was a consideration while working on technical solutions. Their product, Nanoclean AC Filters, are designed to complement the filtering screens of wall mounted ACs to enable them to remove PM2.5 and other pollutants present in the air indoors while also cooling it—all at an affordable cost. Thus, these filters turn any AC into an air purifier. This reduces the cost significantly as there is no recurrent energy cost, yet it is effective and hassle free.

Nanoclean Global has received a global patent for Nasofilters and has filed patents for more than five prototypes. The company was honoured by the Indian National Academy of Engineering (INAE) and awarded the INAE Young Innovator & Entrepreneur Award in 2018.

Frugal innovations by university start-ups like Nanoclean Global stand to gain in many ways, such as scientific excellence of the team, access to laboratory facilities for continuing experimentation and support of the incubation centre in matters related to business coaching, IP protection and market strategies. Such innovations not only help keep us safe, but also contribute to the growth of our economy.

Light a Million Lamps—Chary's IKYA ReD

Have you ever replaced a fused tube light and wondered if it could be used in some way or was throwing it away the only option? The same question crossed the mind of Narasimha Chary, a student of class VII, when he saw a mound of discarded tube lights in a junkyard in his village in Telangana. He asked the local electrician this question and was promptly dismissed.

But Chary couldn't let go of this question, and kept researching and tinkering with ideas and instruments, and wires and tools even after he graduated in electronics from an open university. His education gave him a better understanding of the mechanisms, but it was his inquisitive mind, perseverance and determination that kept him deeply engaged for over a decade to find a solution to this problem.

He broke down the problem for himself. The essential components of a tube light are a choke, starter and filaments. Once the filaments open up and are not able to generate a certain voltage, the tube fails to light up. Chary realized that even when a tube light is fused, it contains about 5 milligrams (mg) of mercury, which is sufficient to burn for about 180 hours. He tried different ways to generate light, even without the filaments, choke and starter. Finally, he tried using an electrical circuit, housed in a box with pins and connected to each end of the tube. Once this was connected to the socket, the tube lit up again and continued to burn till the mercury ran out. This simple and low-maintenance IC-integrated tube light does not need filaments, induction coil, starter or choke and is an energy saver. A truly frugal innovation, it costs about Rs 100 per light, compared with Rs 200–500 for a new tube light.

Besides the simplicity and affordability of the technology, Chary's formula is also environment-friendly. Discarded tube lights with a substantial amount of residual mercury are a potential source of pollution, as they contaminate air, water and soil. Chary's patented technology, IKYA ReD, has helped light up more than 10 lakh tube lights across the country.[2]

In this case, the solution was not unknown to technically trained people but no one really thought of applying it on a large scale to reuse fused tube lights. There could be many possible reasons for this. Perhaps, it was not attractive enough as a good business option because of the shorter lifespan of the repaired tube lights or cumbersome to be adopted at the household level. Though Chary's innovation is somewhat limited in terms of its scope, in suburban and rural areas, where people use lights for shorter lengths of time than in urban areas, its adoption was successful.

However, when a similar problem of utilizing unspent energy in laptop batteries was tackled by a team of IBM India, it led to the discovery that 70 per cent of discarded Li-ion laptop batteries have enough energy to power an LED light for at least 4 hours per day for a year. This led to the development of UrJar, a rechargeable lighting device. With refinement of design and technology, UrJar now not only offers a cleaner and cheaper energy option by reusing discarded laptop batteries but also helps minimize e-waste, which is a big problem in this electronic era.

Water

The SDGs have identified water as key to achieving inclusive sustainable development. Not only as goal six (Clean Water and Sanitation) but water also remains at the core of many other SDGs. Even SDG 2, which deals with zero hunger and, hence, sustainable agricultural productivity, depends on the availability of safe water.

Eco-friendly Waste-water Treatment for Safe Use in Agriculture

Agriculture accounts for 70–90 per cent consumption of the world's freshwater resources. India, with 4 per cent of the world's water resources, supports 18 per cent of the global population. Of its total water resources, about 80 per cent is used for agriculture alone, which ensures food security to nearly 1.3 billion people.[3] Growing population, rapid urbanization and industrial growth are all mounting pressure on water resources, leading to increasing water scarcity and forcing resource-poor farmers to use waste water. Globally, nearly 80 crore farmers practice waste-water agriculture, India being the third largest practitioner of waste-water agriculture after China and Mexico.[4] This poses a serious threat to the environment and human health. However, use of treated waste water presents a viable option to sustain agriculture.

Waste water used for agricultural irrigation is of different types[5], including one or a combination of the following:

- domestic effluents consisting of black water (excreta, urine and associated sludge) and greywater (kitchen and bathroom waste water)
- effluents from commercial establishments, including hospitals
- industrial effluents of different kinds
- storm water and other run-off

Greywater is an important component of water conservation. It comprises 50–80 per cent of residential waste water and offers great potential for irrigation purposes. Treated waste water is subjected to one or more physical, chemical and biological processes to reduce its harmful effect on plants and overall health hazards. However, the available technologies were mostly borrowed from other countries and hence either too costly or ineffective in the long run. A technology developed by ICAR for waste-water treatment is frugal and rugged, that is, it works smoothly for over eight years and meets the National Green Tribunal (NGT) Act, 2010, guidelines for agriculture use.

Scientists from ICAR–IARI's Water Technology Centre worked towards designing a cost-effective and efficient system of waste-water treatment suitable for agriculture use. Models for different situations, ranging from 1500 litres per day (LPD) to 1,00,000 LPD capacity and requiring as little space as 2 square metre per kilo litre (sq m/KL) of water treatment were standardized. The system requires minimum power consumption for water flow, primarily making use of physical gradients created by moderate slopes and gravels. The aquatic reeds used in this system are mainly *Phragmites communis* (common reed) and *Typha latifolia* (broadleaf cattail), which have the ability to release high levels of oxygen from their root system that helps microorganisms to colonize and thrive better in the root zone. These microorganisms clean up the water by degrading most of its organic contaminants and absorbing some of the inorganic effluents. The foliage of reeds can be harvested once every year, leaving the roots and stubble intact to grow again. The dried reeds are used for making particle board and briquettes/pellets for clean sources of energy. This model, besides making water safe for irrigation, reduces the use of fertilizers (nitrogen and phosphorus), and leads to higher crop yield and additional employment generation.[6] This innovation does not boast new technology, but optimizes the best combination of available knowledge to suit prevailing conditions. The treatment cost reduces by 50–65 per cent without compromising on the quality, lowers pathogen load by 99 per cent, is practically zero-energy and 1500 times more sustainable than existing practices. This model of waste-water treatment is gaining popularity in different parts of the country.

Safe Drinking Water

India's Ministry of Water Resources has estimated the country's current water requirements to be around 1100 billion cubic metres per year, which are estimated to rise to 1447 billion cubic metres by 2050. The Asian Development Bank has forecast that by 2030, India will have a water deficit of 50 per cent. Polluted rivers, inadequate water harvesting and recharging of reservoirs, poor storage—India stores only about 6 per cent of its average annual rainfall of 1170 mm—and leaky sewerage infrastructure are recognized as the main contributors to this serious condition of water deficit. It is feared that this water deficit will be unmanageable in the future if not addressed quickly.[7] Though the government and departments concerned are taking measures to meet this enormous challenge, there are several examples of individuals and NGOs contributing with their innovations, technology development and ground-level dissemination. According to NITI Aayog, nearly 40 per cent of India's water supply comes from groundwater sources. But with aquifers polluted by household, industrial and agricultural waste, and biological contaminants, there is real danger of the spread of waterborne diseases like diarrhoea, cholera and typhoid. Besides, there is a serious challenge of groundwater depletion. Many S&T practitioners and start-ups look at these challenges as opportunities to work on innovative solutions.

Many frugal innovations and methods have been devised in recent years to provide sustainable solutions to tackle water scarcity. These are results of science-based, simple to use and affordable solutions for problems faced at various levels.

The problem of fetching potable water from distant sources for household purposes by BOP users prompted Ramesh Kumar and Swathy Ravindran from IIT Madras to innovate a low-cost and sturdy water transporter, Roll Pure, which not only provides easy transportation but also purifies water.[8] While rolling, biocidal material kills biological contaminants and debris is filtered by a pressure differential created by a pump bellow. The young innovators were awarded BIRAC's Students Innovations for Translation & Advancement of Research Explorations–Gandhian Young Technological Innovation Award Grant (SITARE–GYTI), presented at Festival of Innovation (FOIN) event (see chapter 7 for details).

Many of these innovative technologies are implemented by people's participation, as part of corporate social responsibility (CSR) by business groups or by deploying modern technology tools to address various dimensions of the problem. Taraltec Disinfection Reactor is one such frugal innovation for decontamination of groundwater. This device was developed by Anjan Mukherjee, a former marine engineer. Taraltec Disinfection Reactor, a 'fit and forget' device, converts contaminated water from borewells and handpumps at the source to clean potable water by killing 99 per cent of the microbes present in it. The device converts the kinetic energy of the fluid into millions of targeted micro-bubbles, each acting as a localized reactor generating heat, pressure and turbulence that release intense energy packets during the collapse of bubbles and kill microbes. The device can be retrofitted to a handpump or a motorized borewell in roughly 30 minutes without any special training. The pumping of withdrawing water from the handpump powers the device without any additional energy and there is practically no maintenance cost.

Thus far, several hundreds of units have been fitted in rural areas with CSR support by major corporate houses, including, Reliance Industries, Tata Trusts, WaterAid and Lupin Human Welfare and Research Foundation. This is estimated to have impacted the lives of more than two lakh people.

Piramal Sarvajal is an interesting CSR initiative by a corporate group to design and deploy innovative technology solutions and business practices for creating affordable access to safe drinking water in underserved areas. They are reaching out to more than 5 lakh consumers daily through water-purification plants, water ATMs, etc., across twenty states. Likewise, several passionate and impactful social enterprises such as Swajal, Uravu Labs, Vassar Labs, WEGoT and Indra Water are deploying technologies or unique operational models that make a difference. If such interest to address serious issues sustains, we will ensure a safe future for our progeny.

Ekam Eco Solutions: A Journey in Making an Impact

Every year, a city like New Delhi uses about 20 per cent of its water consumption for sanitation. Not only is this a non-judicious use of water, but the treatment of the volume of sewage created is a daunting task. The limited waste-water treatment facilities are incapable of handling such large volumes. Almost 80 per cent of sewage is discharged untreated into the rivers. The waste water is not recycled either.[9]

In this scenario, Ekam Eco, a start-up incubated at IIT Delhi, came up with a practical solution by developing Zerodor Waterless Urinals. This commercial product is an example of 'innovate and make in India'. In these urinals, a simple ball valve replaces the liquid sealants and membranes used by other waterless technologies. The hollow low-density polyethylene (LDPE) ball has a density lower than that of common urinals.

It is housed in a cartridge below the urinal pan. At rest, the valve blocks odour from drainage lines.[10] A single installation costs around Rs 7000 inclusive of plumbing costs and saves upwards of 1,00,000 litres of freshwater per urinal installation per year. The life of this product through frugal innovation is upward of ten years.

Ekam Eco has installed over 10,000 Zerodor kits and conserved over 500 million litres of water. But this isn't enough. This experience has helped them launch a suite of other environmentally acceptable products.

Ekam Eco's business has two particular interests—one, to conserve limited freshwater resources by obviating the use of 1 litre or more water to flush about 200 ml of urine; and second, to bring the idea of 'experience' into the design of sanitation systems by making it more welcoming for people to use it. In other words, make it preferable even to the educated middle class, a big segment of society, to use a dry urinal. In fact, keeping washrooms dry, that is not flushing urinals, is actually cleaner than using a flush every time! This is because on mixing of urine and water, the cistern can get furred up and bacteria can grow. Other problems that may occur include water droplets splashing into the air and spreading bacteria or the cistern getting clogged.[11] The Ekam Eco team wants to make sure that it helps the country since the continuous dumping of 80 per cent of our sewage into rivers and using about 20 per cent of our water for toilets every year is making the situation precarious.

Dry Toilets

When the Indian government offered a subsidy under Swachh Bharat Mission to build toilets in every part of the country, several local and foreign tech providers came up with alternative designs and viable solutions suited to different conditions. Ecological

Sanitation (EcoSan) is one such multinational technology provider with experience of promoting both dry-pit and wet double-pit-type toilets in developing countries, particularly in South Asia. Recognizing cultural barriers, EcoSan followed a community-led and PPP-based decentralized approach for construction of dry toilets.[12]

The basic concept of EcoSan is to manage human excreta by decomposing it before disposing. Not only can these organic by-products be handled safely, but these enrich the soil and are beneficial to agriculture without harming the environment. To suit varied geographies and water availability in different parts, EcoSan came up with two-pit toilet designs, which do not have flush systems and consume very little water. The ammonia-rich waste water can be used efficiently for irrigation. Collaborating with local organizations, the teams of EcoSan experts undertake necessary modification of the design of toilets using appropriate technology as per local needs and social norms. These toilets are being used in dry regions of Gujarat, Karnataka, Maharashtra, Andhra Pradesh, Kerala, Odisha, Bihar and Ladakh.

Affordable and effective healthcare is central to the idea of sustainable development. Though most scientific research, technology development and innovations are undertaken at the organizational level, there are some remarkable efforts made by individual researchers who are driven more by passion and purpose than mere professional responsibilities.

Fight Against the Emperor of Maladies

The Vedas proclaimed the importance of a healthy body for happiness, 'sarve bhavantu sukhinah, sarve santu niramaya'. Though by maintaining overall cleanliness, occurrence of many infectious and communicable diseases can be minimized, no matter how careful we are, there will be certain life-threatening diseases for which neither the cause nor a complete cure is known. Cancer, a disease characterized by uncontrolled growth of cells in any part of the body, is one such example. It is dreaded as much for its lack of cure as the high cost of treatment.

Though there has been evidence of cancer as a disease since ancient times, the focus it received over the past fifty years has been due to its spread across all age groups, geography and social strata, as well as growing awareness about the disease. Compounded by the fact that there are about 200 different types of cancers to which no single cause can be assigned, there is lack of preventive measures and low efficacy of curative measures.

Breast cancer is one of the most common cancers affecting a large number of people, with 16,71,000 new cases[13] being reported every year globally. India accounts for nearly 10 per cent of these cases.[14] One of the most effective means to contain breast cancer is through surgical interventions like mastectomy. Though a lifesaver, it leaves a patient with negative feelings about their appearance and can lead to deep mental trauma.

Throat cancer is more prevalent among the economically weaker section (almost 80 per cent), widely addicted to chewing tobacco. As patients mostly seek medical help only after reaching stage three or stage four, their voice box gets completely damaged and patients have to undergo laryngectomy, leaving them voiceless.

The concern and imagination of two separate doctors led to the development of an innovative post-surgery solutions for patients of breast and throat cancers. Another researcher developed a diagnostic tool for timely and accurate diagnosis of cancer. Affordability, ruggedness, indigenous technology and ease of application of these frugal innovations promise sustainability of these technologies.

Sampoorti–Poorti

Dr Pawan Mehrotra, an ace biochemist, has trained in some of the best universities and overseas laboratories. While researching the cancer biology of breast cancer in India, he was moved by the embarrassment, taboo and emotional distress that women face after undergoing mastectomy. The other revelation was that there was no significant attempt to devise a holistic post-mastectomy solution that would allow patients to regain confidence and dignity after this serious physical loss. The problem was addressed only sporadically through import of expensive and ill-fitting external breast prosthesis or, in a few cases, through very expensive implant-based breast conservation surgeries. Post-surgical complications included back pain, abnormal gait and postural abnormalities. This unmet social need prompted Dr Mehrotra to look for a solution.

Various limitations and sceptics did not deter Dr Mehrotra from forming Aarna Biomedical, a social enterprise he believed was the vehicle that would keep his focus on development and sustainable commercialization for social impact.

The journey included learnings in design and engineering under the tutelage of Prof. P.V.M. Rao at IIT Delhi; BIG Scheme support from BIRAC for prototyping; development space and support by Translational Health Science and Technology Institute (THSTI)–DBT; investment by Social Alpha of Tata Trusts; and engagement with several NGOs and hospitals. Resorting to a user-driven, inclusive and innovation approach, Dr Mehrotra's team aimed to develop a 'post-mastectomy kit' that would provide a holistic solution. It comprises one external pre-made silicone (medical-grade and CE-certified[15] material) breast prosthesis (available in different sizes and shapes as per the choice of patient), two pocketed brassieres in which the prosthesis can be worn (available in different sizes and colours), two prosthesis covers, one prosthesis holder, a manual and an outer waterproof kit to discreetly accommodate all the components. The development involved product design, material selection, procurement, manufacturing and packaging with usual challenges in each case. A noteworthy step involved indigenous development of a semi-manual setup to mix, degas and dispense silicone instead of an elaborate metered dispensing unit. Aside from addressing problems like bubble formation in silicone formulation, this local machine costs one-tenth of the imported metered dispensing units. Dr Mehrotra says there are no manufacturers of 'lightweight silicone gel' breast prosthesis in India. Aarna Biomedical has cleared necessary trials and validations in various settings in different parts of the country.[16]

Aarna Biomedical developed an affordable silicone breast prosthesis called Poorti, which is a post-mastectomy kit to provide a non-invasive and simple solution to survivors of breast cancer who have lost either one breast or both breasts completely as part of the surgical intervention. Sampoorti allows for the creation of women entrepreneurs who offer a trial to the patients serving a cause and earning a value in return.

True to the names 'Poorti' and 'Sampoorti', which in Hindi literally mean fulfilling (a need) and complete fulfilment, the company already caters to markets in different parts of India, as well as to patients in Nepal, Bangladesh, Sri Lanka, Russia and Africa.

Aum

Dr U.S. Vishal Rao has been a witness to the post-surgery trauma of many patients with throat cancer in India. Often, patients belonging to the low-income group cannot afford a costly voice-box device, which costs around Rs 20,000–45,000.[17] So, he set out to develop a low-cost and innovative device that would be effective as well as affordable for BOP patients who have lost their ability to speak due to laryngectomy.

Supported by BIRAC funding of the DBT, after years of experimentation, tests and trials, he made a device that can be placed through the patient's neck and connected to the food pipe, where it functions as a voice box. In about two months' time, the patient gets used to the device and can speak. Named Aum, the silicone device weighs just about 25 grams and costs less than Rs 75. Each unit is expected to last for about six months, just like imported models.[18]

Together with his friend Shashank Mahesh, Dr Rao established Innaumation Medical Devices Private Limited in February 2019, a start-up for efficient diffusion of his existing and forthcoming innovative devices.

It is heartening to find this start-up spreading the spirit of 'science to serve' through its new initiative, the Innaumation Medical Grant—Design Health Thinking. Through this initiative, the start-up supports young researchers and potential innovators, especially medical, pharmacy and engineering students, with a small funding of Rs 25,000 to Rs 50,000 to solve local problems with innovative solutions rather than depending on costly imported alternatives. The company believes that the goal of knowledge is service, hence profiteering is not the goal. It is an excellent step forward in encouraging science-backed frugal innovation in health services.

OncoDiscover

Early detection is key to the containment of cancer and recovery. There are a number of research institutions backing start-ups that are engaged in developing kits for detection of specific types of cancer at various stages. A common objective of all the start-ups is to provide affordable and reliable diagnosis at an early stage of disease. DBT's BIRAC is the main funding agency sponsoring such studies, whereas public or private research institutions incubate such innovators.

Cancer spreads in the body through a process known as metastasis, wherein cancerous cells break away from the primary tumour into the bloodstream and are transported to various organs. These cells, known as circulating tumour cells (CTCs), act as seeds for the growth of new cancers and are responsible for over 90 per cent of patient deaths.

Because CTCs are as low as one cell among 1000000–10000000 leukocytes[19] of the peripheral blood of cancer patients, detecting a CTC is like searching for a needle in a haystack.

Liquid biopsy is the common name given to a set of blood tests that detect various biomarkers such as CTCs, circulating tumour DNA (CtDNA), exosomes, circulating free DNA (CfDNA), etc. These tests are minimally invasive and require drawing of blood, unlike a solid-tissue biopsy, which is invasive and painful. A liquid biopsy may be used to detect cancer at an early stage, plan treatment, determine how well the treatment is working and if the cancer has returned. Being able to take multiple samples of blood over time may also help doctors understand what kind of molecular changes are taking place in a tumour.

The only test approved by the US Food and Drug Administration (FDA) for detection of CTCs is CellSearch. Actorius Innovations and Research, an R & D start-up co-founded by Dr Jayant Khandare and Aravindan Vasudevan in 2013 at NCL's Venture Centre and initially funded under BIRAC's BIG Scheme, has developed OncoDiscover. It is a liquid biopsy technology and a patented polymeric-multicomponent system which is an in-vitro diagnostic tool developed in India that can rapidly isolate, detect and enumerate CTCs from a cancer patient's blood with high precision and sensitivity compared with CellSearch, even though both techniques work on a similar antibody-based affinity mechanism.

OncoDiscover is a useful tool for detection of progression and relapse of cancer. The test is minimally invasive as it requires only 5 ml of a patient's blood and can rapidly and sensitively detect CTCs. The results of this test can be used by oncologists for understanding important variables, such as occurrence of CTCs in blood, response to anticancer therapy in patients and potentially the outcome of the disease with time, etc. Patients who have been treated for a primary cancer will find this test useful for early detection of cancer recurrence, which is a major reason for cancer-based mortality. Clinical outcomes of this test conducted at Tata Memorial Hospital, Mumbai, have been accepted for publication at prestigious scientific conferences, including the ones organized by the American Society of Clinical Oncology (ASCO) and the American Association of Cancer Research (AACR).

CTC analysis for cancer outcome and its management in patients is already approved and well characterized in the US, China and Europe. The CDSCO, a regulatory body for drugs and medical devices in India, has granted a licence for manufacture and sale of OncoDiscover. Actorius Innovations has priced the OncoDiscover liquid biopsy at Rs 15,000 per test, which is more affordable than CellSearch, which costs close to Rs 1,00,000 per test.[20] This frugal but scientifically important innovation is efficient and holds the potential for widespread use around the world.

Agriculture

Affordability and availability of appropriate technologies and quality inputs are key to sustainability in agriculture. Farmers today are faced with more challenges than before. One major factor contributing to this is the unpredictability of changing climate. The other is degradation of natural resources, such as soil and water. To mitigate and manage these challenges, farmers need to apply all their ingenuity 'to produce more from less for more'. It is, therefore, no wonder that one comes across the largest number of frugal innovations in agriculture. Some of these are purely developed by the farming community, whereas others are either devised by research organizations or commercial suppliers of seeds, fertilizers, pesticides, and irrigation infrastructure and farm machineries.

Many innovations made by farmers are to reduce the workload or increase productivity. The largest number of farm innovations therefore are in some form of mechanization or automation of existing implements. Poor roads, scarcity of water and shrinking farm labour have led to the development of a range of automated field devices, like an auto-driven water-lifting system for irrigation using the principle of traditional system of yetha by Manjegowda from Mandya, Karnataka, and the low-cost windmill for water pumping by Assam's Hussain and Ahmad.

There are many other instances of frugal innovations, from primary processing of perishable farm produce to low-cost biopesticides and biofertilizers, growth boosters, and many more, which are devised by scientists and simple enough to use by farmers, even though some of these may not have direct market values. The seed cubes devised by scientists at the Forest College and Research Institute in Mettupalayam, is one such example. The technique of using mudballs containing seeds and dispersing them in forest lands or farm fields with the onset of the rainy season is nothing new. It was popularized by organic farming pioneer Masanobu Fukuoka, the author of *The One-Straw Revolution*. However, in practice, a large number of mud balls crack, scattering seeds that dry up and fail to germinate. Jawahar, a young innovator from Mettupalayam, developed seed cubes in three simple steps. Seeds of trees like neem, pungam and tamarind are first primed with locally available plant-boosting materials. The planting medium is prepared using a mix of soil, vermicompost, sawdust, arbuscular mycorrhiza (a microorganism) and guar gum. A simple wooden frame, which looks like a bigger ice-cube tray, is used to make cubic sections on this layer from this wet-soil mix. Finally, one or two seeds are put in each of these cubes while they are still moist. Once dry, the cubes are extruded. These cubes are easy to transport in cardboard boxes to planting site, are less prone to shattering than seed balls and the soil mix supports the seeds so that they germinate well.[21] Seed cubes are particularly useful in organic farming, afforestation and creating urban forests—a concept drawing much interest to improve air quality in residential neighbourhoods.[22]

At the national level, institutes under the ICAR and the State Agricultural Universities (SAUs) develop technologies to address the problems of national and regional importance. The ICAR launched the National Agricultural Innovation Project (NAIP) 2006–2014, with the key objective of laying greater emphasis on innovation in agriculture and developing sustainable value chains by involving all stakeholders from different sectors in a consortium mode. It supported a wide range of projects, from upstream basic innovation to the development of frugal technology. Close to 300 technologies were developed under this project, of which about 100 were commercialized for wider reach. The NAIP also provided an agribusiness platform in the form of Agrinnovate, a unit of the ICAR specifically created to facilitate commercialization of innovative agriculture technologies developed by ICAR research institutes and SAUs.

Recognizing the effectiveness of open innovation programmes, the Ministry of Agriculture, Government of India, in partnership with Startup India, launched the Agriculture Grand Challenge. The ministry identified twelve problem statements and sought solutions from the start-up ecosystem. After a series of workshops, multiple screenings and interviews, twenty innovations were finalized. Each innovation clearly defined the critical gap and presented a solution. The nature of the innovation ranged from mobile-based apps to rapid soil-testing kits and handling of agricultural commodities in the National Agriculture Market (e-NAM).[23] A few examples given below amply demonstrate the diversity:

- **A Simplified, Sensor-based Rapid Soil-health Testing Method to Test Macro and Micronutrients:** To achieve best productivity by using the right and the correct amount of fertilizers required, soil health tests need to be done regularly. The Government of India launched a Soil Health Card Scheme, under which farmers were given free soil health cards and the cost of the tests was subsidized. However, the turnaround time for the results was long. Therefore, the agriculture ministry looked for a solution that ensured rapid and accurate tests. Soumya Rao, a young soil scientist from Bhubaneswar, created a prototype of a rapid soil-health testing kit. The kit completes the test in about 15 minutes and costs less than Rs 5000.

- **Real-time Assaying and a Quick Grading-solution for e-NAM to Handle Agricultural Commodities:** The e-NAM is a virtual market with a physical market (mandi) at the back end. The portal networks existing mandis to create a unified national market for agricultural commodities for pan-Indian electronic trading. The e-NAM portal tackles the problem of information asymmetry and ensures that farmers get the best price for their produce. Innovative technologies, which help grade agri-produce very quickly, are linked to get better prices.

- **Mobile-based Applications:** There are more than 53 crore smartphone users in India.[24] Therefore, the numerous mobile-based apps launched by government departments and commercial agencies are a good example of frugal innovation in agriculture since they offer smart, cheap and effective solutions to farmers' problems. Developed by the Centre for Development of Advanced Computing (C-DAC) and Department of Agriculture and Cooperation (DAC), the government uses Kisan Suvidha to provide information regarding weather, market prices in mandis, retail sources to purchase seeds, fertilizers, farm implements, etc. It was launched by Prime Minister Narendra Modi in 2016. Launched by the Indian Farmers and Fertilizers Cooperative (IFFCO) in 2015, IFFCO–Kisan Agriculture provides agriculture advisory, weather, market prices, kisan credit card (KCC) services and agriculture information. Krishi Mitr–RML Farmer is another app that has information about 1300 mandi prices, use of fertilizers and pesticides, information on 450 crop varieties, weather information at 3500 locations across 50,000 villages in seventeen states. Pusa Krishi, developed by IARI scientists in 2016, not only provides information about resource conservation technologies and farm machinery developed by the institute, but also by other ICAR institutes. Developed by Tata Consultancy Services (TCS), mKRISHI is another successful app that provides a range of personalized services to farmers on their mobile phones in local languages, such as agri-advisory, best technologies, pest alerts, weather forecasts, and market and supply management (see chapter 5 for details). Krishi Gyan Apps, besides providing general information on farming, connects farmers directly with experts to get the right advice. Crop Insurance, on the other hand, helps farmers calculate insurance premiums in simple steps, provides information about the cut-off dates, sum insured, subsidy, etc., and is linked to the ICAR web portal, catering to all stakeholders. These apps reduce farmers' vulnerability by alerting them against many biotic (damage due to diseases and pests) and abiotic (drought, excess/uncertain or sporadic rains, high and low temperature conditions) stress factors inherent to farming. In the wake of the economic recession experienced in the recent times by almost all economies to uneven extents, rapid depletion of natural resources, air and water pollution, challenges of climate change and natural calamities, frugal innovation offers a powerful option for sustainability of our ecosystem. Built around the intrinsic philosophy of minimum use of easily available resources, improved functionality and maximum value for every rupee spent by the consumer, these find a big market share and, hence, make good business sense. Thus, in today's context, the concept of frugal innovation, is equally relevant both in developing and developed economies. The world has come to realize the importance of a frugal mindset, which can be leveraged to instil robustness, portability and mega-scale production, and other conditions for developing frugal innovations,[25] and these are only going to be adopted more in the coming years in a variety of fields. As we race towards a future replete with cutting-edge technologies, frugal innovations gain greater salience in offering eco-sustainable solutions.

5

Creating Value through Frugal Innovations

In the modern world of business, it is useless to be a creative original thinker unless you can also sell what you create.

—David Ogilvy

Innovation and Beyond

Once a pathbreaking innovation is made, it is only logical to expect that an efficient mechanism will help spread the technology for the benefit of society at large. Be it an upgraded mobile phone with user-friendly features at a lower price, a car that reduces fuel consumption or technology that processes waste and reduces environment pollution, the users always benefit. It is through this adoption and commercialization that innovators recover the money spent on development and make a reasonable profit for their own sustenance and funding future projects.

This also stands true for institutional programmes. Be it a public-funded research institution or an R & D facility in the corporate sector, the returns from an investment largely determine future investments. While planning the distribution of resources, a programme that holds a promise of viable product development with commercial value invariably gets priority and more fund support. In most academic institutions in developed economies and some top institutes in India, researchers are encouraged to pursue innovative ideas and products that come out of these and can be licensed by faculty and students to the industry. They can even use these to start their own ventures. This system not only helps generate more funds from royalties of successful products, but also provides the most effective means of diffusion.

Innovation diffusion involves reaching out to members of various communities and enabling them to get access to the innovation. Besides the utility and nature of the innovation, local needs and conventions, communication channels, institutional arrangements, need of specialized skills, time and social contexts exert a significant impact on diffusion. Even if a product addresses the need of a group of customers, such as Tata Nano and Chotukool, its diffusion is not guaranteed. On the other hand, products like Vortex's Gramateller and GE's portable ultrasound machines are sold in many countries.

Often, frugal innovations tend to be grassroot innovations and relevant only to a small number of users, compared with other types of frugal innovations. In a study covering fifty innovations in the healthcare sector in different countries,[1] it was found that entrepreneurs at the grassroots target a specific gap and solve local challenges by developing products that require in-depth knowledge about processes in local ecosystems. Also, such entrepreneurs lack the necessary infrastructure or business expertise to scale up their solutions.

Customer Value-based Solutions

To ensure the success of a frugal innovation, a customer-value-based approach is essential to enable easy acceptance among customers. Developing a product that fits the needs of a certain group of customers is not enough. The firm needs to ensure that a complete solution meets the requirements, the context, and is compatible with the local business environment and stakeholders, including regulators. In Mexico, cement producer CEMEX launched the Patrimonio Hoy programme so that poor people could build their own houses. It created a building club of a few members who would take joint responsibility for making payments and homebuilders could add one room at a time. Customers applied as a group of three to the local Patrimonio Hoy cell and committed to a 70-week membership, wherein they would remit a small weekly payment (about Rs 740–1110) to be held as credit towards the housing material that would be delivered to them by CEMEX. By changing the mode of distribution, CEMEX was able to eliminate several middlemen and thus reduce the cost. To encourage the programme, promoters and CEMEX regional cell managers of the Patrimonio Hoy programme were compensated on group repayment performance and the length of commitment to the programme.

This particular frugal innovation was a success because knowledge and resources were made available to customers, the quality of the product was ensured and the mobilization of community participation enabled repayment of loans to take place on time. Therefore, every link in the chain needs to be thought through and suitable strategies devised. At times, there are invisible forces that deter customer from acquiring a frugal product. For example, the middlemen in the case of Patrimonio Hoy and firms promoting frugal innovations need to study the situation from the customers' point of view to interpret the real reason for lack of participation.

To ensure easy acceptance by customers, the context of users and the roles of partners need to be planned carefully. A case in point is Meesho, an Indian-origin social commerce platform founded by IIT Delhi graduates Vidit Aatrey and Sanjeev Barnwal in December 2015. It enables small businesses and individuals to start their online stores via social channels such as WhatsApp, Facebook, Instagram, etc.

The mission of Meesho, short for 'meri shop' is to focus on growing the number of resellers. By June 2021, it had acquired more than 1.7 crore of these resellers.[2] It was an incidental meeting with a few owners of small businesses who had discovered the power of social media platforms that led to the inception of Meesho. The founders realized that the potential of e-commerce could be leveraged only if suitable value-added services were offered to the sellers on the Meesho platform instead of just connecting the buyers with the sellers.

Meesho provides resellers with a customer relations management (CRM) feature to drive sales growth. Every time a customer visits an online store, the reseller gets a notification. The ability to chat with a potential customer gives the buyer some room for bargaining, which is normally not possible on other e-commerce marketplaces. Meesho's business model is largely built around small and medium businesses, and most of the sales are processed within a trust network or repeat buyers. The ability to chat helps resellers identify trusted buyers within their network, even though the buyer and seller may not have interacted before. For every transaction, a reseller makes a commission of 10–20 per cent.

Why is Meesho successful? Not many aspiring entrepreneurs are able to invest in building a separate website for their products and in such situations, online marketplaces like Meesho come into play. The platform enables housewives, students, professionals and others to start their stores/boutiques. Most importantly, Meesho has enabled interactions in local languages rather than just English since a majority of Indians are not conversant in English.

The example of Meesho shows that for easy acceptance by users, a complete solution needs to be offered instead of just a standalone product. While any traditional software service firm would offer the solution at a steep fee to users, Meesho took the plunge and created a solution that worked on commissions on customers' orders instead of asking for a hefty fee for use of its software platform. This approach made the proposition very attractive for people to set up their shops on the Meesho platform.

Innovation Diffusion

Innovation diffusion determines how well the innovation is adopted by consumers. Studies have shown that a superior technology, ease of adoption, overall acceptability and social compatibility play a major role.

In case of frugal innovation, diffusion depends on maximizing the value created for society rather than market profitability. For rapid diffusion of frugal innovation, approaches such as diversification, scaling out, scaling deep and scaling up are found particularly effective. Diversification of products and services helps serve more needs of consumers, whereas sharing the innovation with others helps reach more people. Diversification can involve adding on complementary products and services. For example, a microfinance firm can add micro-insurance products to offer complete solutions to SHGs. Scaling out is an approach wherein a successful diffusion model in one market is replicated in other markets.

Scaling up involves changes in policies, procedures and laws to ensure easier diffusion. Scaling deep involves changing beliefs and cultural values for easier acceptance of innovations.[3] These approaches can be useful, particularly in cases where communities resist the adoption of a frugal innovation owing to age-old beliefs, superstitions and prevalent cultural norms. At times, innovators may need to act as social transformers. To maintain commitment towards the social cause, social entrepreneurs need to listen to diverse voices. It is a good idea to include representatives of beneficiaries in planning a diffusion strategy for frugal innovation in the beginning itself so that their concerns are duly addressed.[4]

The success of Vortex's Gramateller is one such example. The Gramateller is an automated teller machine (ATM) targeted at rural markets. It consumes only a fraction of the electricity compared to traditional ATM machines and runs on open-source software, thus operating at a much lower cost. Further, the Gramateller enables fingerprint-based biometric authentication, which is a boon in societies where the number of illiterate ATM users is quite high. Based on this core concept, the Government of India has launched biometric-authentication-based door-to-door money transactions by linking post offices with the banking system.[5] This disruptive idea promises successful diffusion, particularly in rural India where illiteracy is high and easy access to banking is lacking.

Business Strategies to Help Diffusion of Innovations

The contexts of frugal innovations are varied, implying that there is no single strategy for effective diffusion. To develop a strategy, the four-As approach may work better than the usual four-Ps (product, price, promotion and place) approach. The four-As are:

- awareness
- access
- affordability
- availability

There are several reasons why the four-As approach may be more effective than the four-Ps approach. Consumers who wish to benefit from frugal innovations may not become aware of the innovation through traditional channels of communication. This is because literacy levels among BOP customers is lower and they may not be exposed to promotions shown on premium-priced television channels. Similarly, making upfront payment for a product like a solar lantern may be a challenge for BOP customers.

Let's consider what British Petroleum (BP) did to develop a biomass-pellet-based smokeless stove for underprivileged consumers in India. Traditionally, rural women collect sticks, shrub, grass and other biomass materials, and sometimes cow dung, to be used as fuel for cooking. They spend roughly 2 to 3 hours a day on gathering fuel and direct burning of biomass materials produces acrid smoke. The pollution is worse when the fire is lit inside the living shelter. Finding a solution was not easy, as cooking habits differed from place to place.

The company first recorded videos of the cooking process in different areas and analysed them to interpret the issues that were important for BOP customers. The objective was to build a modern, smokeless and easy-to-use stove that would also be an aspirational product for the target segment. Most products targeting BOP customers typically focus on the functionality but do not identify the emotional needs of these customers. Therefore, the product needed to follow global safety standards and yet be affordable. Accordingly, the company made pellets from agriculture waste that served as fuel for the stove, that was aptly name Oorja. Priced at Rs 675, it had a chamber for burning pellets, and a mini fan powered by rechargeable batteries and controlled by a regulator that fans the flames. The company worked hard to first determine a price point that would be suitable for the target consumers, and then worked backwards to arrive at the cost. Thus, they used the formula of 'price minus profit equals cost' and not the more traditional formula of 'cost plus profit equals price'.

Marketing is another important aspect of innovation diffusion. Firms must carefully choose the product features that are highlighted in their promotions. The Nokia 1100 was primarily targeted at BOP customers in rural markets and its powerful torchlight was a great way to promote it, as large parts of rural India suffer from power shortages. This was equally true in Africa, but the success there was also attributed to the voice messaging feature, which addressed the problem of illiteracy.

The way a product is distributed can also have a major impact. In order to reach the BOP customers in rural areas, Unilever in India began appointing women as distributors of their products, including fast-moving consumer goods (FMCGs) like soaps, shampoos, toothpastes and other household items. Typically, these women, called Shakti Amma, belonged to poor families from nearby localities. In order to ensure affordability, Hindustan Unilever Limited launched their products in small sachets and priced them very low. The company claims that this initiative boosted the confidence of the women involved, resulted in increased participation in household decision-making and augmented their sense of self-worth. Other unique techniques include distribution through channels that are normally used to sell other products. For example, the mobile telephony industry in India has enjoyed high penetration by making recharge vouchers available at grocery shops and other utility stores.

While attempting to include the aspirations of customers in product design, it is also essential to define the boundaries of the innovation. If the product is perceived as 'cheap' and offered to economically disadvantaged customers in a manner that is perceived as 'condescension', even a well-designed product can fail in the market. This is probably the reason why the Tata Nano, a well-engineered car that was offered at very low cost to customers, failed to win the hearts of buyers. The global press dubbed it the $2000-car, equivalent to Rs 1,00,000 at the time of launch, and the tag of 'cheap car' possibly deterred many buyers. A car is considered a status symbol in many countries and customers do not wish to be regarded as 'poor' by society despite being the owner of a car. Hence, the branding of frugal innovations must ensure that the social image of the innovation is consistent with the customer's social aspirations. In fact, Tata Motors branded the Tata Ace, a minitruck, as an upgrade from a three-wheeled goods carrier.

The culture of a society has a huge impact on innovation adoption and diffusion. While many studies have focused on a 'national' culture determining the diffusion patterns of innovations, the culture within a country, e.g., India, can differ significantly among the different demographic segments and geographic regions. Individualism is an aspect that significantly affects the pattern. In cultures characterized by individualism, the autonomy of individuals is higher and trials for new products may be quite encouraging for firms. However, a study of BOP customers revealed that collectivism trumps individualism in poor societies and individual autonomy is curbed. This has an adverse impact on diffusion, and the approach to be adopted for collectivist cultures involves taking community members into confidence during the diffusion process. Further, societies that are characterized by high masculinity necessitate that male members of the community take the lead role in the innovation diffusion process. Several firms in emerging markets, especially rural and BOP communities, rope in the elders of the community, typically men holding high positions in society, to testify for the innovation.[6]

The types of risks that negatively impact product adoption include: social risk (how will the reference groups perceive the product adoption by users); physical risk (the possibility of physical harm caused by the product's adoption); functional risk (uncertainty about the availability of complementary or supplementary products); and switching costs (the perceived costs that users will have to incur to switch to the new product). Hence, a strong collaboration is desired between technology providers, innovators and marketing firm, particularly to assess physical and functional risks. Firms can rope in trusted laboratories and research agencies to certify the usage of products that customers may be reluctant to use.[7]

Cultural beliefs and mindsets of people often cause resistance to new ways and practices. In spite of the availability of several options of mechanization for precision farming, agri-advisories and custom hiring, small and marginal land holders, constituting over 70 per cent farmers in India, face serious challenges when it comes to finding affordable labour to perform farming operations. This is not because of lack of technology but they either do not have proper information or access to technology. Even for most of those who do have the knowledge, it is not economically viable to adopt modern farming technology. Taking advantage of this gap, technology providers are coming up with innovative and situation-based solutions. Kamal Kisan[8] is one such solution provider, based on a pay-for-use model. In its list are standard machineries as well as many locally designed and manufactured implements, such as vegetable planters, diggers, backpack weeders, etc., much needed by small cultivators.

Sometimes, people's changing perspectives, and government policies and schemes can also contribute towards the diffusion of innovations that were available for a long time but not adopted. Increased use of the Happy Seeder in recent times is one such example. Burning of paddy stubble (residue) after harvest, in order to prepare the land in time for sowing wheat, is a regular practice followed by farmers in the northern Indo-Gangetic plains. In order to prepare the field and complete sowing of wheat in time, farmers opt to burn the residue. This not only is a major contributor to severe air pollution in the NCR and adjoining territories, but also has an adverse impact on soil health. Happy Seeder is a technology based on the principle of 'zero tillage' and used to sow wheat without burning rice residue. The Happy Seeder is a tractor-mounted machine that cuts and lifts rice straw, sows wheat into the bare soil and deposits the straw over the sown area as mulch. In this way, the wheat seed can be planted without getting jammed by the rice straw.[9] This technology is eco-friendly and saves time and water. Though the technology was available and tested for over ten years, it received attention only since 2018–2019, when air pollution in NCR became a national issue and state governments announced up to 50 per cent subsidy on it.

Experimentation is essential to ensure that innovation meets the requirements of customers. In the case of the biomass stove, the design had to be altered several times through the inclusion of new features, such as a stainless-steel sleeve with ceramic coating that could withstand the heat and yet was easy to clean. A network of partners helped in fulfilling the requirements of product design and diffusion. Collaboration with the IISc helped gain access to the faculty who experimented and improved fuel efficiency, while a tie-up with NGOs gave access to village-level entrepreneurs called jyoti ammas, who helped push the product into homes.[10] Better coordination between different departments, academia, industry partners and NGOs working at the grassroots, such as Honey Bee Network, is needed to identify gaps, find innovative solutions, test their acceptability and expedite marketing using appropriate business models.

A good business model ensures that all stakeholders appreciate the value being offered, market segments are accurately identified and the revenue generation mechanism is clearly specified. Customers evaluate the benefits delivered vis-à-vis the costs that they have to incur. Therefore, while planning the business model, focus needs to be on value creation for customers.

The choice of the right business model can make the difference between failure and success. One of the key reasons behind failure of a business model is the perpetuation of myths about BOP customers. The belief that BOP customers only prefer cheap products and are not amenable to adoption of new technologies may not necessarily be true. Failure of products like the Tata Nano bear testimony to unfounded myths about BOP customers; it being necessary to align affordability with the consumers' context. SELCO, a rural renewable energy service firm in India, has worked on dispelling myths and implementing a suitable business model to enable BOP customers gain access to renewable energy solutions. During the last three decades, SELCO has installed close to 5,00,000 solar solutions for BOP customers across India. SELCO's successful business model has many distinguishing features, such as recruiting local youth, offering doorstep services through energy service centres; and earning a modest profit to ensure sustainability of operations and financing solutions that match the cash flows of BOP customers. SELCO believes in creating a social impact instead of focusing only on financial returns, and diversity among the employees, in terms of gender, caste, colour, religion, etc., which helps reinforce the trust, empathy and sensitivity between customers and SELCO. By following the right business model, SELCO has been successful in diffusing technology that has positively impacted more than 10 lakh people.

Business Models for Driving Success with Frugal Innovations

Innovative business models, like asset-light models, which depend heavily on outsourcing or renting can help drive down costs while also enhancing efficiency. Since resource constraint is an inevitable aspect of frugal innovations, asset sharing and usage-based pricing mechanisms are beneficial. The use of optimization technologies, e.g., cloud technologies, big data, etc., help in bringing accurate and cost-effective innovations. If the objective is to enable many people to access a product that they are currently unable to access, suitable technologies can be borrowed. The launch of Nollywood by Kenneth Nnebue enabled millions of Africans to gain access to locally made video content using existing technology (VHS tapes) and an innovative business model (straight to video).

Tattvan E-clinics, set up by Ayush Mishra with headquarters in Pune, have a team of full-time doctors and nurses available to ensure quality consultations through videoconferencing and telemedicine facility. The centres also host doctors of global repute to conduct one-on-one follow-ups of out-patient departments (OPDs) for better diagnosis and treatment. Patients need to visit a hospital only for surgeries. For all other ailments, they can do a consultation from their city. Because of this, patients not only save on travel, boarding and lodging, but even the consultation fees charged at Tattvan E-clinics are substantially lower than physical visits. The Tattvan E-clinics have already spread or are in the process of doing so to various countries, like

Afghanistan, Iraq, Bangladesh, Tajikistan and Uganda. The example of Tattvan E-clinics shows the importance of business models in diffusion. Such facilities are valued more at remote places and during unusual times, when normal movements of people are grossly restricted due to calamities such as floods, landslides, epidemics or wartime.

To enable better diffusion, firms opt for partnerships with other stakeholders for greater penetration in the intended segment. Since firms that are for-profit have different goals, structures, values and cultures from not-for-profit partners, they build a common platform to leverage each other's competence. The partnership between Save the Children, an NGO, and Procter & Gamble (P&G) is a case in point. This partnership has enabled P&G to reach out to children in developing countries with its sanitary products. In order to test frugal innovations in real-market conditions, partnerships during the testing process can help. Through such partnerships, P&G has been able to undertake testing of innovative products with the right consumers without having to deploy expensive resources.

In a similar vein, several banks have partnered with NGOs to offer microcredit to BOP customers living in marginalized communities. For example, Drishtee Foundation is an organization that has set up 1600 customer-service points and is offering banking services to millions of people through its tie-up with the State Bank of India (SBI). Drishtee's reach in rural communities has helped it offer microloans to BOP customers for SHG-promoted businesses, farming, rural housing, insurance products, etc. Encouraged by its success, several similar partnerships exist now.

Another effective model to promote, inter alia, frugal innovations is to put in place dedicated structures, e.g., subsidiaries, that provide customers a value proposition that may not be in line with those of the parent.[11] These structures and networks address all issues related to innovative offerings or expansion of business to large budgetary segments. The launch of the 'Ginger' brand of budget hotels by the Indian Hotels Company Limited (Taj group of hotels) as a separate subsidiary (Roots Corporation) is one such example, which succeeded after testing several prototypes. A more prominent example from the hospitality industry would include the Ibis hotel chain from Accor, a large global group. The affordability was a result of efficient processes, redesigned systems and avoidance of unnecessary frills.

After successful market testing of an innovation, partnerships with other group companies can help in diffusion. An example of this is mKRISHI. There was a huge need to help farmers who, while at the mercy of vagaries of monsoon, were averse to the use of technology-based solutions. Lack of access to farm inputs, technology, information, credit and marketing support hinders small farmers' efforts to improve their income. In this scenario, mKRISHI uses information and communication technology to help solve farming challenges. The beauty of mKRISHI lies in the fact that the technical team continues to innovate and upgrade the app to retain its relevance. It delivers a range of personalized services, such as agro advisory, best practices, alerts, weather forecasts, and supply-chain management to farmers in local languages on their mobile phones. Farmers have formed groups of 1000 under the Progressive Rural Integrated Digital Enterprise (PRIDE) initiative of TCS. This enables the 300-odd PRIDE groups to order farm equipment through bulk deals, thereby bring down costs. mKRISHI makes use of social networks, mobility, analytics, cloud, and Internet of Things (IoT) to undertake predictive analytics for crop acreage and yield, crop health, soil status, weather and pest forecasts. While TCS has been the technology provider, the diffusion is being managed by Tata

Trusts, which, inter alia, is engaged in raising agricultural productivity and reducing rural poverty. In this case, TCS leveraged the network of Tata Trusts for easy diffusion of the innovation. The market diffusion was also helped by another group firm, Rallis, which deals in agri-inputs and engages with farmers. The mKRISHI digital platform has helped more than 5.5 lakh farmers get easy access to weather forecasting, query solving, and marketing of agri-products.

Offering customers the opportunity to test the innovation for a trial period is another important feature that helps assure the value being offered through first-hand experience. Hence, to promote frugal innovations, particularly in the fields of agriculture, medicine, education and financial services in rural areas, trials should be made available to customers either free of cost or at a nominal cost, involving local networking partners.

As discussed in an earlier chapter, a market-driving approach also contributes significantly to successful adoption of innovations. The starting point for a market-driving approach is a detailed study of the various players present in an industry and the business models that are prevalent. A case in point is the agri-business division of ITC, a firm predominantly focused on the FMCG industry in India. Farmers in India have traditionally depended on middlemen in various agricultural markets for selling their crops. This dependence has often resulted in unfair treatment of farmers with respect to the prices paid to them for their produce. ITC realized that lack of market knowledge and the inability of farmers to gain access to the wholesale traders in agricultural markets resulted in this situation.

ITC realized that it could play a key role in helping farmers get better prices for their crops. In June 2000, the group launched e-Choupal. It enabled ITC to get access to cheaper and more reliable supply chains, while creating a price advantage that could be passed on to customers through frugal innovations. The firm's willingness to support new ideas and experimentation helped foster diffusion. In the case of ITC, the entrepreneurial orientation exhibited by the senior management of its agri-business division and the chairperson of ITC, were vital for the diffusion of e-Choupal, which now connects more than 40 lakh farmers across India. What is noteworthy is that instead of trying to remove the current intermediaries in the chain, e-Choupal makes use of their physical transmission capabilities, in aggregation, logistics, counter-party risk and bridge financing, but isolates them from the chain of information flow and market signals. The farmers manage village Internet kiosks and get easy access to information in their local language on weather and prevalent market prices, disseminate knowledge on scientific farm practices, facilitate the sale of farm inputs and purchase farm produce from other farmers' doorsteps. In this way, the farmers can gauge the market demand and take a decision on the right time to sell their produce at the appropriate price. The system also enables farmers from various locations to aggregate their purchases of farm inputs, e.g., seeds, fertilizers, etc., and thereby ensure lower prices for each farmer. In addition, e-Choupal, also helps farmers with solutions to their specific problems, if any, from expert scientists through e-consultations. This is a prime example of leveraging complementary resources to create more value. In order to gain access to complementary resources, firms need to build partnerships and extend the network of firms with whom they have built relationships.[12]

To ensure a customer-orientated approach, firms need to be willing to adjust and modify the features of their products as per market needs. This can be quite challenging at times, considering the fact that firms invest in setting up large factories with built-in capacities. In order to cope with this, manufacturers have been adopting collaborative approaches, often referred to as 'industrial symbiosis' or 'industrial ecology', wherein traditionally disparate industries join hands to share materials, energy and other by-products. The objective of frugal innovation is to make products more affordable and flexible in the manufacturing process along with lower-cost materials in the supply chain that can help to ensure low prices.

Context-oriented Business Models

Various case studies based on frugal innovations targeting BOP customers suggest that the first step should be to understand whether a technological solution can help fulfil an unmet need.

The development and diffusion of a pro-poor toilet system by Dr Bindeshwar Pathak, founder of NGO Sulabh, holds important lessons for frugal innovators. The Sulabh toilet model looks similar to others but differs quite significantly. The waste from the toilets is directed into one of two pits dug near the toilet. When one of the pits is full, a valve enables the waste to be directed to the other pit. The dry waste from the first pit can be used as manure in farms. This toilet model fulfils the goal of a sustainable sanitation system. However, changing the habits of communities is not an easy task. In order to spread the message and educate users, marketers need to compete with television shows and other attractions.

For example, to drive the mission to eradicate open defecation in India, the government launched the Swachh Bharat Mission. Celebrities like actor Amitabh Bachchan have featured in many television messages on Doordarshan exhorting the public to refrain from open defecation and to adopt sanitary toilets that the government has helped people to build.

Another firm, Samagra, founded by Swapnil Chaturvedi, aims 'to enable the urban poor to lead healthier, productive, dignified and empowered lives'.[13] Toilets can have a huge impact on human dignity, and Samagra realized that a nation's growth and health is directly impacted by access, or the lack of it, to clean toilets. Diarrhoea kills more than 2,00,000 children each year in India and a major cause of such diseases is lack of access to clean toilets. So, it uses information and communication technologies, and human-centred designs to make sanitation services accessible, affordable and acceptable to the poor. The success of Samagra can be attributed to the passion that Chaturvedi carries for the venture. Chaturvedi, who worked in the US from 2001 until 2009, left his cushy job as a software engineer and enrolled in a master's programme in design and management course in Northwestern University, Illinois, to help him embark on a new career in social change. He researched sanitation issues affecting the urban poor in developing countries and began working on designs and business models that could be applicable to India.

Chaturvedi's business card reads, 'Poop Guy—I love shit'. In March 2013, he, along with his wife Tania, founded Samagra Sanitation, a hybrid social enterprise entity that has both a for-profit (Samagra Waste Management Pvt. Ltd.) and a non-profit (Samagra Empowerment Foundation) arm working at the intersection of design, technology and behavioural science to tackle open defecation. To use Samagra's toilets, slum-dwellers pay a monthly fee of Rs 75 per family. They receive an ID card that can be used at any Samagra toilet, any number of times. The toilets are managed as Samagra franchisees by individuals from low-income communities, which also include three Samagra sanyoginis (local women), who receive 100 per cent of the revenue collected, which they use for cleaning and maintenance of toilets through a team of trained cleaners. Samagra also makes sure it engages with users beyond just toilets. It has taken efforts to build higher engagement through innovative approaches. The LooRewards platform created by the company uses community toilets as channels for user engagement to promote hygienic behaviour. It does this by rewarding slum residents for using toilets regularly, on-time payment and performing tasks such as washing of hands. For example, one of the rewards is a 15 per cent discount on low-cost sanitary napkins. Samagra also tied-up with SBI to provide banking services to the communities living in slum areas

Experimentation has been a key reason for the success of Samagra. In August 2010, Chaturvedi travelled to India from the US to test his concepts and discovered in the first few days that most of the ideas developed in the classroom were not sustainable on the ground. He dropped out of the university and began a series of experiments between 2011 and early 2013, including a small pilot project in Raipur, constructing toilets that didn't require plumbing, and had to be cleaned manually (this project ended because it did not comply with the Prohibition of Employment as Manual Scavengers and their Rehabilitation Act, 2013), and a poop-to-power project in Bhubaneswar, which too wound up because it lacked financial sustainability. His operations in Delhi and Bengaluru failed as well. He was almost ready to give up when a friend invited him to Pune.

However, these failed experiments eventually led to a more well-rounded business model that involved the community and the government. Through previous experiments, they learnt that context is important while designing solutions. It is necessary to identify the pain points of customers and formulate suitable engagement strategies like LooRewards to gain traction. Samagra's success can be attributed to the choice of partners, such as the Bill and Melinda Gates foundation, Government of India's Swachh Bharat Abhiyan and Ashoka, an organization that supports the world's leading social entrepreneurs. By virtue of becoming an Ashoka fellow, Samagra gained access to an expansive network of peers and partners.

The commercialization strategy for a frugal innovation should be to focus on the outcome that the customer is expecting from the innovation. It needs to be understood that for customers belonging to BOP markets, every addition to the cost ends up making the purchase incrementally unattractive. After analysing the situation from the customers' point of view, innovators can also choose to involve the customers in the process of producing the product, which can help reduce the cost, and they can pass on the benefit to the customers. Consider the example of Rasna. In the 1980s, Rasna launched the soft drink concentrate in packs that were priced very low. The customer had to add sugar and water to the concentrate to make the drink. For customers who could not afford to pay for the ready-to-drink but expensive lemonades or fruit juices available off the shelf, Rasna was an excellent choice. The company had carefully broken down the value components and taken an innovative step in getting the customer to co-create the product at a fraction of the cost of the popular competitors like Thums Up and Limca. This concept, known as co-creation, may not work for affluent and time-starved customers, but for BOP customers it may also give them pride in contributing to the value that they will consume in future.

The role of leaders in innovation teams is often crucial in enabling the success of frugal innovations, more so in case of breakthroughs than innovations that build a little on other innovations. The leader nurtures a culture that allows for candid debates and discussions and prepares the team to learn from many 'productive failures'. This is how everyone is encouraged to continue experimentation till the product or idea becomes acceptable to the target group. The first task of the leader is to be able to see possibilities that others have missed. Most businesses end up copying competitors and trying to push their products through competitive pricing strategies. What leaders should focus on instead are strategies like 'blue ocean strategy', which reduce and eliminate standard features in a product while raising or creating others that can help differentiate the offering.[14] Leaders can drive frugal innovations by reimagining the product, the value chain, the complementary services, and the total cost for the customer.

Frugal innovations require firms to create an ecosystem that includes R & D alliances. Governance assumes high importance to ensure that the rights and interests of various partners are well protected. At CISCO, the Cisco Hyper-Innovation Living Labs (CHILL) is working on different problems faced across the globe (e.g., the future of work, digitized supply chain, etc.). In 2020, CISCO merged the CHILL accelerator with its global innovation centre. CHILL has taken the approach of making a team's discoveries available to all participants in proportion to their investments. In this way, it ensures fairness and distributive justice towards the contributors whose innovations may be used later by other groups.[15]

One of the successful innovations in the field of healthcare is Aravind Eye Hospital, which was set up in 1976. It has brought down costs by following a cross-subsidization model and the hospital is able to offer eye surgeries to poor patients free of cost. In this case, the value chain was reinvented. The intraocular lenses that had to be imported at a high cost were manufactured at a much lower cost through the hospital's own subsidiary—Aurolab. Doctors were assisted by a very competent team of nurses to ensure high productivity. By performing more surgeries every day, the cost incurred by the hospital was reduced. The outreach of the hospital was extended to several nearby towns and villages through eye camps and by setting up more hospitals. Sponsorships were also received from local NGOs and organizations like Lions Club and Rotary Club. In the case of Aravind, Dr Venkataswamy, affectionately referred to as Dr V, was a retired eye surgeon who led from the front and inspired his colleagues to fulfil his dream of eradicating needless blindness. He showed that the right business model can help achieve social goals.

By working in proximity with members of society, it is possible to develop frugal innovations that can help fulfil the dreams of BOP customers. One such example is the firm Masisa, which offers customized wooden furniture in several South American countries like Brazil, Chile, Argentina and Venezuela. The firm believes in adopting a socially inclusive approach to improve the lives of its customers and other citizens. It works vertically with several plantations to source wood board and solid wood products. To kick off this project, several stakeholders were roped in Curitiba, Brazil, for developing do-it-yourself (DIY) furniture. Its partnership with Placas Centro helped train carpenters produce improved furniture at low cost, while women from low-income group families were trained as sales agents. The carpenters thus trained can also become micro-entrepreneurs, whereby the goal of social inclusiveness can be fulfilled.

Similarly, the case of Bandhan, a non-banking microfinance firm which later obtained a banking licence, exemplifies a well-meaning partnership between the bank and the borrowers. About two decades ago, Bandhan, presently known as Bandhan Bank, was founded with the explicit goal of helping the poorest of people to escape poverty through microloans. The members of partner groups included only those families that did not have any able-bodied male, owned less than 0.2 acres of land, had children engaged in jobs and not receiving education, and for whom informal labour was the main source of income. The selected members were included in a skilling programme to ensure that the chosen borrowers were well-equipped to undertake farm or non-farm-based business, like livestock farming. A small stipend was paid to each family every week so that household expenses were taken care of and the women could focus their energies on growing of the business. Bandhan representatives met the borrowers each week, keeping track of the progress in their respective businesses and providing necessary advice that helped them grow their business.

Enabling Shared Value for Stakeholders

In a world that is witnessing increased contention over scarce resources while the divide between the rich and the poor widens, it is not only sensible, but also makes good business sense to take a step back and work towards shared values for all. Several firms have taken the lead by embedding a social mission in the corporate culture and developing suitable innovations that deliver value to various stakeholders. Firms engaged in the food business often highlight nutrition as their marketing strategy while creating innovative offerings. The approach involves understanding underlying social conditions and barriers that hinder progress. For example, Nestlé developed Maggi Masala-ae-Magic, a micronutrient-reinforced spice mix, aimed at children. This product is tasty, affordable and within easy reach of BOP families, with minimal nutritional value. Thanks to the extensive marketing network of Nestlé and its strategy to be distributed by social organizations like Drishtee that are working towards social and economic development among BOP segments, it became a commercial success.[16]

Shared value is crucial for frugal innovations, and relationships between firms and customers need to be based on exchange of value and not only a product/service. To ensure this, the value needs to be assessed on the basis of economic, social and ecological parameters. Social inclusiveness, environmental protection and economic resilience can create sustainable business models. Commercial firms have been facing the brunt of society's ire due to unbridled consumerism that puts profits above everything else. In this milieu, there is growing support for creating shared value by enhancing the competitiveness of firms while simultaneously advancing economic and social conditions.[17] Firms need to use resources in a conscientious manner and keep the interests of diverse stakeholders in focus while making business decisions. Frugal innovations promote mindful consumption and thereby fulfil the tenet of sustainability by caring for self, community and nature.

Sustainability as a goal is challenging, and regulations and compliances may get more stringent in times to come. However, compliance can be viewed as an opportunity for innovation and the value chain can be made more efficient through better carbon management, lower energy consumption, use of sustainable raw materials and clean energy. Corporate houses can fulfil their commitment towards CSR by supporting innovative technologies in line with sustainable practices. More importantly, issues of sustainability are well-recognized from the point of view of frugal innovation, and its practice needs to be ensured through appropriate steps. Not just raw materials, even value creation and post production activities ought to meet the sustainability criteria. The complete cycle of a product needs to be planned and managed carefully, and science-based viable technologies developed at every step.[18] Generally, reductions in carbon footprint can be achieved by reducing the emission of greenhouse gases, increasing the mix of renewable energy, reducing consumption of water in products and agriculture, reducing, reusing and recycling plastic used in packaging, reducing paper consumption and undertaking sustainable sourcing of raw materials like palm oil, tea, fruit, vegetables, sugar, soy beans, etc. If corporate firms focus on diffusion of frugal innovations, partnerships with tech innovators and/or systemic research to develop frugal solutions, it can also provide a fillip to their sustainability efforts.

Though frugal innovations were earlier aimed at BOP customers, it is no longer easy to distinguish between BOP customers and the mass market. Therefore, the philosophy of frugal innovations can become a mainstream idea and fulfil the needs of customers across demographic segments while ensuring profitability for firms. Research has shown that many examples given by C.K. Prahalad for businesses targeting BOP customers are applicable to mass markets as well.[19] Therefore, more and more firms are launching frugal innovations that fulfil unmet needs of the mainstream customer and promote the cause of sustainability. For example, Mobycy is India's first electric scooter and bike sharing app that provides last-mile connectivity at a low cost and without any hassles. The entire process of renting has been automated and access to bikes has been made very easy for commuters alighting from trains or buses. The app is available to everyone and is useful even for customers for whom cost is not of the utmost consideration. While society expects the government to provide low-cost transportation for all citizens wherever it is required, there is dearth of easy-to-access and low-cost transport facilities for those alighting at bus stops or metro stations in India. Mobycy is fulfilling the needs of society and consumers while being environmentally friendly and affordable. Further, Mobycy helps to reduce the strain on government resources and thereby helps in creating greater value for the community at large.

The no-frills approach in the design of frugal innovations helps to fulfil the tenets of sustainability. A big need for BOP communities living in rural areas is access to cheap and dependable transportation. Traditional modes of personalized transport options like taxis are too costly for most rural people.

In order to provide affordable transport, Ola has launched bike taxis in rural and semi-urban areas. The Ola bike service is currently present in 200 cities and towns, of which 80 per cent have a population of less than 10,00,000 people. The agriculture ministry in India is planning to launch an app for locating heavy farm machinery like tractors, land levellers, harvesters, Happy Seeder, etc., on hire-and-use mode.

Enabling Diffusion through Partnerships

Even after a frugal innovation has been developed, the team driving it needs to build partnerships with various agencies for diffusion. The role of the government and other agencies is to help alleviate the risks associated with the innovation and help in propagating its use.[20] The government must also protect the interests of the underprivileged in society. The concern that customers have is not always unfounded, as in several instances, particularly, BOP customers get cheated by unvalidated technology, are forced to purchase unnecessary items or exploited by unscrupulous parties promising high financial returns. Necessary instruments need to be built into government policies and programmes to safeguard BOP customers from being exploited. Capitalist enterprises can sometimes increase their profits at the expense of such customers who are unable to move out of their poverty trap.[21]

In this regard, the government or agencies appointed by the government need to play the role of a watchdog. For example, cheaper garments produced in large-scale factories by multinationals can easily crowd out the local handloom products painstakingly created by artisans. The Indian government has been offering support to artisans through the textile ministry and help to boost exports and support the sale of goods through exhibitions in various cities in India and abroad. Perhaps the government can infuse the philosophy of frugal innovations in sectors like handicrafts to make these more competitive in terms of price and sustainability initiatives, e.g., use of more eco-friendly materials in production and promotions through digital channels. Firms like Fabindia[22] have developed supply chains comprising local artisans from the BOP while creating products targeted at higher-income customers. Thus, Fabindia can play an important social role through its business model. It is India's largest private platform for products made using traditional techniques, skills and hand-based processes, linking over 55,000 craft-based rural producers to modern urban markets. This not only creates a base for skilled artisans and sustainable rural employment, but also preserves India's traditional handicrafts in the process. Its whistle-blower policy helps Fabindia monitor practices from the perspective of ethics and fairness.

Close associations between industry and public institutions (educational or research-focused) can foster frugal innovations. When governments promote entrepreneurship, economic and social aspects can derive the benefit. Diffusion is enhanced through common language, physical and technological proximities. In countries where legal systems are weak or underdeveloped, or insiders favoured at the expense of outsiders, the economic efficiency is bound to suffer. Internal factors like lack of human capital (education and managerial skills), resources (financial capital and information), and networking capabilities also impact the success of frugal innovations.

The knowledge of managers and skill level of workers are crucial too. The ability of a firm to build trusted relationships with external networks and social connections influence the success of innovations. The government's policies for supporting foreign direct investment (FDI) and technology transfer can foster absorption by local firms through the formation of clusters. In fact, technology transfer can compensate for lack of human resources in host countries.

Institutional distance and void have a major influence on the fate of frugal innovations and their diffusion. Companies in developing countries dominated by BOP customers have significantly different challenges from those in developed nations. Multinational firms, in particular, face tough challenges targeting BOP customers. The presence of trustworthy institutions gives confidence to multinational firms, while reducing transaction costs and uncertainty of laws and regulations. The example of Samagra illustrates that innovators need to build partnerships with government agencies, e.g., local municipal corporation in the case of Samagra, for successful deployment. Supportive government policies can encourage foreign investments in technology development. Given a level playing field, multinational companies, with their advanced scientific capabilities and partnerships can play a vital role in introduction and diffusion of frugal innovation.

The role of the social entrepreneur needs special attention from the perspective of shared value. If a business is to ensure shared value, it must create goodwill among the stakeholders. Social entrepreneurs engage in entrepreneurial activities with the goal of fulfilling a social mission. If 'caring' is imperative for a social entrepreneur, then all business issues will be looked at with this lens of 'caring'. The recognition of the business opportunity will then be regarded as 'caring about' while value creation will become 'care giving' and the exchange will be 'care receiving'. The caring attitude of entrepreneurs cascades down to the employees and the partners of the firm. This helps in promoting inclusiveness not only inside the firm but more importantly in marginalized sections of society and customers belonging to BOP communities.

There are examples of not-for-profit initiatives being started as developmental programmes addressing the needs of the BOP, which in the time course became full-fledged social enterprises. Kilimo Salama, meaning 'safe farming' in Swahili, a crop insurance scheme, promoted by the Syngenta Foundation for Sustainable Agriculture in 2013 turned into ACRE Africa, which is the brand name of Agriculture and Climate Risk Enterprise Ltd., the largest private sector index-based insurance programme in Kenya and Africa. It links farmers to insurance products that will both protect them from losses suffered due to failed crops and allow them to confidently invest in their farms. Partnering with leading insurance companies, by 2017 it covered more than 16 lakh farmers in Kenya, Rwanda and Tanzania, and generated around Rs 55 crore in gross premium. ACRE is one of the first agricultural insurance programmes worldwide to reach smallholders using mobile technologies. Support from Global Index Insurance Facility (GIIF), a dedicated programme of the World Bank Group, was vital.

Entrepreneurial experimentation characterizes how entrepreneurs use their knowledge, networks and resources through experiments with new technologies. In 2003, the social innovation unit of Vodafone won an innovation grant from UK's Department for International Development and M-Pesa was born out of a pilot project to provide the BOP access to financial services through simple mobile phones. A prototype was prepared and tried out with 500 microfinance borrowers in two locations in Kenya—Nairobi and Thika. The dealers were trained to operate the payment application and provide customer care, and the pilot clients were given a free mobile phone with the software pre-installed in it. Suitable modifications were carried out, and the final product was a generic mobile payment service.

Thereafter, Safaricom, Kenya's mobile network operator, has opened up the application programming interfaces (APIs) of M-Pesa to allow third parties to develop and integrate their applications, and there is a complete integration of M-Pesa with the country's banking and payment system. The regulator, the Central Bank of Kenya, has worked closely with Safaricom on product functionality, legal compliance, technical requirements, prudential controls and consumer protection. Safaricom was able to win over the regulators by showcasing the product as a social initiative to fulfil the financial inclusion agenda of the government. Thus, the interplay between political power and the dynamics that govern a frugal innovation have a major impact on the shared value. Governments can foster propagation of frugal innovations by documenting them in the form of case studies and distributing the knowledge through common platforms.

Active support of governments can usher in a wave of frugal innovations. For example, the Manipur government has been promoting bamboo shoot products to encourage sustainable local practices. The state government has recognized it as an industry with low investment and high employment potential in the region. Workers have been trained in teamwork and efficient working procedures, and the output is encouraging. Government agencies like the National Bamboo Mission and Department of Agriculture & Farmers Welfare in the Ministry of Agriculture have come forward to support the project. Banks have provided loans while the Manipur University and the Central Agricultural University in Manipur have offered scientific backstopping.[23]

Good Practices Promote Innovation

The Government of India and the NITI Aayog, a policy think tank of the Government of India, have published a volume of case studies titled 'Good practices resource book' that documents how technology and good governance can help solve problems using frugal approaches. The Honey Bee Network (honeybee.org) also serves to document frugal innovations from across India and over the last twenty-five years has resulted in the documentation of over 1,00,000 frugal innovations. The Government of India has been making efforts to help entrepreneurs wishing to reach out to partners through the website agnii.gov.in, which is managed by the Office of the Principal Scientific Adviser to the Government of India. The website also provides guidance on IPR and IPR facilitation for start-ups.

The example of Taipadar, a tribal village in Chhattisgarh in India, is also an example of how shared value needs to be created. Though a stream named Ganeshbahar, which later on goes to meet the Godavari river, flows through the village, Taipadar remains dry for eight to nine months a year. Punit Singh, a socially conscious scientist from the IISc, was inspired by E.F. Schumacher's book *Small is Beautiful*. During his PhD at the Karlsruhe Institute of Technology in Germany, Singh took up a project on environmentally sustainable solutions like turbine pumps to generate electricity for local use. Singh applied this knowledge to solve the water problems of Taipadar by trying out a ram-pump, which uses kinetic energy of the water to pump it up and does not require electricity. The objective was to make sure there was water throughout the year for drinking and irrigation. However, the ram-pump, which was working with a makeshift dam made up of sand bags was not able to fulfil the needs of the local villagers. To create a permanent solution, Singh built a case for a turbine that would not only help provide enough water but also generate electricity for the village. The local government officials were approached with a proposal wherein KSB, a German pump manufacturer, was willing to supply the pumps and even fund the project. Understandably, Singh became a bit of a local hero and is now being approached by other villages for similar solutions.

One of the immediate policy-level interventions that is required is to strengthen the IPR regime. This would create better incentive mechanisms for frugal innovators to graduate from innovating on processes to focusing on outcomes. A simplification of the existing patent system would lower the threshold for one to seek intellectual rights protection and create enough incentive for oneself and social welfare. While domestic enterprises would migrate from frugal process innovations to frugal outcome innovations, a more mature regime would encourage multinational corporations (MNCs) to increase their focus on emerging economies as their lead markets for frugal innovations. Another imperative at the policy level is to build on native knowledge and ingenuity. Most grassroots-level frugal innovators attempt to solve their personal problems by applying local ingenuity, but since these approaches are ad hoc, they are not stored for later use. One way of preserving and exercising native knowledge and ingenuity is by setting up rural innovation laboratories where quick and frugal proof-of-concepts can be conducted to test out new ideas. These labs can be set up on a hire basis, where people with ideas can economically collaborate with others, including academic institutions and industry.

Another intervention can be in the context of corporate philanthropy, where firms have traditionally donated towards social welfare activities like orphanages, schools, hospitals, etc. Though implicit in the spirit of CSR,[24] a more nuanced articulation to foster frugal innovations may provide a strong fillip to institutional partnerships towards development and diffusion of the same.

Looking ahead, it is evident that customers will become more value-conscious and therefore innovations will have to ensure the creation and delivery of suitable value to the intended customers. There are three things that will drive customer value in the future. First, increased digitalization suggests that getting customers to engage in more co-creation of value will help companies drive down costs while offering personalized solutions to customers. Second, business models will need to be socially oriented, hence, the social impact of the innovation will have to be more relevant. The problems faced by poor consumers can provide the triggers for innovations. The proposed SSR Policy, 2019, of the Government of India proposes to build-in a sense of social responsibility in every scientific research undertaken by the public or private sector, and hence, promote socially relevant innovations. Third, shared value models like the sharing economy will gain more takers and peer-to-peer sharing of resources will become more common. While several industries like hospitality (Airbnb) and cabs (Uber) have already been transformed, others like banking are going through a metamorphosis. Digital peer-to-peer lending platforms (e.g.,www.faircent.com) have built easy-to-use features for both lenders and borrowers.

The silver lining in all the examples mentioned above is that they will put a smile on the face of the customer, thanks to the enhancement in the customer value!

6

Market-driven Solutions

Marketers influence demand by making the product appropriate,
attractive, affordable, and easily available to target consumers.[1]

—Philip Kotler

The market potential for frugal innovations has been a well-acknowledged fact given the huge gap in resources among the well-off consumers and the billions of not-so-well-off consumers around the world. However, the number of successful frugal innovations are only a handful and that raises the question—what does it take to create products and services that would succeed in serving the needs of resource-constrained customers?

The Creation Process of Frugal Innovations

The creation of a product to meet a specific need is not a challenging task for firms engaged in the particular industry of which the product is a part. The challenge is to create a good quality product at a low cost so that customers belonging to BOP communities can afford it. The complexity of product designs and the accuracy levels required to ensure good quality are unfortunately not feasible at a low cost. The R & D activities need to ensure adherence to exact values to reduce the margins of uncertainty. To meet safety specifications, any innovative product needs to undergo well-controlled experiments. This reduces uncertainty and ensures good repeatability of operational performance. Accurate models in engineering and materials can help widen the range of size, shape and type of parts for standardization to ensure lower costs for fabrication and assembly. To ensure sustainability, the chosen technologies must result in minimized waste generation and lower emissions. Modern manufacturing processes like near-net shape manufacturing, 3-D printing, etc., can be utilized to build frugal products with minimal waste.[2]

In recent times, several approaches have been adopted by firms to ensure reduction of waste, reuse of materials and equipment, and use of automation for reducing costs. In the construction industry, the use of prefabricated building blocks has helped reduce construction waste. Likewise, bricks are made from fly ash instead of clay. Fly ash is a by-product in thermal power plants and is otherwise known to be a pollutant. The use of waste materials eliminates the process of disposal of that material and also uses a resource available at a very low cost and (hopefully) in abundance. In this way, the cost of the product can be reduced.

The literature on innovations is predominantly focused on product innovations. The stories on radical innovations have focused more on products (e.g., Tata Nano) than on services. Services are events (or a series of events) that are delivered to customers (often in real time and with the active participation of the customer). In India, several low-cost service innovations like the Aravind Eye Care System have shown the effectiveness and power of frugal innovations. The Aravind Eye Care Hospital performs over 4.6 lakh operations every year and this scale enables it to offer free surgeries to the poor and charge a reasonable fee to those who can afford to pay. In order to create a service-oriented innovation, it is essential to start by understanding the context, that is, the circumstances under which the event(s) occur and the physical, social, technological conditions that shape the event.[3] It is, therefore, important to define the customer's journey and the experience that needs to be created. To do this effectively, one needs to focus on: cognitive (what people think), physical (how people interact), sensory (what people experience), emotional (how people feel) and social (how people share) issues.[4] Service innovations need to focus on customer journeys as there is less clarity on service innovations than product innovations. This calls for firms to adopt a flexible approach while creating and delivering services and creating an ecosystem of trusted partners so that the uncertainties associated with service innovations can be overcome. The capabilities should enable firms to undertake needful automation, proactive personalization, contextual interaction and journey innovations.[5] These become very important while addressing diverse customers in different markets. Automation can help lower costs while enhancing efficiency, proactive personalization can help customers address their specific needs, and contextual interactions help to offer the service as and when it is required by customers and innovations help ease the journey of customers. For example, MedTech [6] has been offering the telehealth facilities by empowering customers to seek personalized healthcare advice through tele-consultation at much lower costs and is a good example of an effective service innovation. To deliver value to customers through service innovations, dynamic capabilities need to ensure flexibilities in service deliveries and co-operation by various service enablers. In order to ensure real-time collaboration among various departments and smooth delivery of service, the use of service champions can help in removing the bottlenecks in operations. These service champions do not owe their allegiance to any particular department and are focused on ensuring unhindered service operations. Whenever they spot a problem that is holding up the service operation, they are quick to address the issue by getting the relevant departments to perform their respective roles.

Customer journeys and event experiences need to be studied in great detail during service creation and delivery. Video recordings of customers' interactions can be analysed to interpret the associated complexities. Learning systems can incorporate stories of customer journeys and the root cause analysis of events' outcomes can be undertaken. The knowledge systems developed while creating and delivering services can help build absorptive and adaptive capabilities of the firm. It is essential to integrate the knowledge assimilated by the firm into decision-making processes that help in the creation and delivery of services.

Firms face challenges in harnessing tacit knowledge from people responsible for service delivery. In order to tap into this knowledge, employees can be incentivized to share their knowledge through meetings and knowledge-sharing sessions. Further, transactional data captured through automated systems (which are driven by self-service technologies) can help decipher the tripping points for customers during service delivery. Firms need to identify the risks perceived by customers and alleviate them through suitable checks and measures. This is more important in the case of services like banking since BOP customers may find it difficult to trust technology-enabled systems.

Owned by BRAC Bank of Bangladesh, bKash offers a good illustration of customer journey tracking in the case of mobile phone-based banking services. It is targeted at poor people who can use the service for cheaper and more efficient cash deposits, cash withdrawals and payment services. Users of bKash do not need to have Internet access on their phones to conduct transactions and only a small fee is levied for the service. In order to ensure that customers are able to transact smoothly, bKash has set up an extensive network of agents across urban and rural areas. These agents hand-hold customers through kiosks set up with the assistance of philanthropic grants offered by the Bill and Melinda Gates Foundation. There are several thousand Union Information and Service Centres (UISCs) across Bangladesh that offer additional services like birth registrations, electricity bill payments, overseas job applications, passport applications, telemedicine, etc. The agents in the kiosks offer technical assistance in case a customer requires it. The Bangladesh government is also supportive of the initiative since the bKash initiative helps promote the welfare of a large number of Bangladeshi citizens.[7]

Therefore, in order to develop a service innovation, a well-planned approach is essential. The service innovation needs to be designed considering the concept that will offer a new solution for customers, the business processes and the value capture model. The value capture model needs to consider the context of customers to ensure that the pricing is affordable. Further, the business partners, communities, NGOs and government agencies can be roped in to help make the service affordable and leverage the expertise/resources of the various partners. The involvement of partners on fair terms is essential to ensure that customer goals are fulfilled. In case the partners feel distanced, they take less interest and the quality of the service delivery suffers. Importantly, the firms need to first analyse their own strengths and weaknesses, and then choose partners that offer complementary skills and expertise.

Consider the case of Solar Ear.[8] Nearly 7 per cent of the world population suffers some degree of hearing impairment and most people have to live without any assisted hearing due to two problems—price of the aid and battery recharging. Targeting this huge need, Solar Ear offers low-cost, solar rechargeable and environment-friendly hearing aids. The firm has partnered with Yunus Social Business and the Lemelson Foundation to support the goals of UNICEF and WHO. Solar Ear hires people with hearing impairments and reinvests the profits to fulfil the social mission, while empowering the workers.

Adapting service innovations to suit the local context is essential for preventing failures. The first step of adaptation should be to figure out the perfect way for interaction with customers. While the use of self-service technologies has helped ensure easy access, the ideal mix of human-computer interactions should be ascertained to ensure customer convenience. Research on service innovations have shown that over-reliance on self-service technologies can lead to customer disengagement. The knowledge sharing for intra-organizational and inter-organizational coordination needs to be effectively managed since variations across geographies, culture, languages, etc., can hinder transfer of knowledge. The compatibility of systems, skills and culture between participating partners is essential for efficiency and seamless coordination.[9]

Partnerships between different organizations with varied capabilities and resources can play a crucial role in the success of frugal innovations. The role of an innovator is to leverage diverse capabilities into a single frugal innovation well-ensconced in an environment within which it has to deliver on its promise. Scientific capabilities and expertise required for an innovation may not be possessed by a single organization and the different capabilities need to be combined to create an innovation.

Let's consider the example of the Trans-African Hydro-Meteorological Observatory (TAHMO), a frugal weather station project in sub-Saharan Africa. Accurate weather forecasting is essential for a continent that is dependent on rain for agriculture and also has important applications in a variety of industries, such as insurance, mining and telecommunications. The density of weather stations in Africa was not enough to offer accurate weather forecasts, which led to considerable damage to crops. The Delft University of Technology, along with the Oregon State University, helped build a low-cost and simple yet robust ground weather station called TAHMO. The set of sensors inside TAHMO measure rainfall, incoming shortwave radiation, wind speed, the direction of wind, barometric pressure, air temperature and relative humidity. The weather station can be installed inside schools and runs on a solar panel the size of a business card. The data is transmitted using 2G and 3G networks. The entire set up requires very little maintenance and has no moving parts. The most interesting aspect of TAHMO is use of data for various business applications, whereby the revenue model has made the rollout of hundreds of weather stations very easy for the authorities. The revenues come from a subscription-based model with the mobile services firm Airtel and a micro-crop insurance firm called ACRE Africa. The total annual cost of maintenance of a station is only $100. The business model has been adapted to suit the local conditions in countries like Kenya, Uganda, Tanzania and Ghana.

In Uruguay, affordable healthcare technologies have been successful in addressing problems faced by local people, such as infant mortality. The human-milk pasteurizer has been successful in eradicating necrotizing enterocolitis (NEC) and decreased nosocomial infections, which in turn has reduced mortality rate among newborns. NEC is a disease that develops when the tissue in the inner lining of the small or large intestine gets damaged and begins to die. This causes the intestine to become inflamed. It typically occurs in newborns who are either premature or otherwise unwell. Symptoms may include poor feeding, bloating, decreased activity, blood in stool or vomiting bile. A nosocomial infection is an infection acquired in a hospital or other healthcare facility. The human milk pasteurizer affords important lessons in frugal innovation. First, to make it more affordable, cutting-edge technology was deliberately not deployed during product design. Second, the reuse of components from other devices was encouraged to ensure easy availability and reduce costs. Third, the milk bank's medical staff also participated in the design process to ensure that all concerns were addressed and the user's point of view was kept in focus. Continuous interactions with users helped in knowledge sharing, particularly tacit knowledge, that could not have been accessed through other means.[10]

Technology road-mapping tools have been found useful to identify resources required to build up innovation architecture, particularly in case of social issues that may be solved using information and communication technology applications. Innovation architecture offers a snapshot that combines the core technologies and scientific knowledge, business systems and the market trends. Blackboard Radio is an AI-powered mobile app that enables children in India to learn English. The approach involves the use of AI to enable children to develop mastery over spoken English through interactive, personalized instructions over a basic smartphone. The system attempts to mimic an English teacher and aims to offer feedback on intonation and expression in the future. As students use the platform for improving their English, the data gathered from students will help fine-tune the content and the instructions by building in safeguards against typical mistakes made by students. The venture has been funded by social enterprise incubator Villgro and aims to improve the English-speaking skills of more than 200 million children across India.

The Japanese firm Fujitsu is engaged in providing solutions to various industries, including healthcare. Recognizing its market potential, Fujitsu has been attempting to find solutions to dementia, which afflicts millions of Japanese people, through open innovation, participant observation and development of technologies based on deep understanding of user needs. Traditionally, Fujitsu has been a business-to-business healthcare company, but for the dementia project, they built new relationships and networks so that a citizen-driven service ecosystem could be developed. Social isolation was soon identified as a major problem affecting dementia patients. Thus, the dementia friendship club was formed to offer an inclusive approach towards these patients.[11]

It is a misconception that stripping down features can perhaps help adapt products to suit the needs of poor customers. Many firms from developed markets have failed to make a headway in emerging markets with this approach. Instead, these firms eventually found that to target emerging markets, reverse innovation could be useful. Harman International, a subsidiary of Samsung Electronics, adopted radical changes to develop products suitable for emerging markets rather than scaling down existing ones. Its approach was to form a team with the software group located in India and the hardware group in China. The senior supervisors realized that achieving the functionality of existing products at half the price and one-third the cost would require a radically different engineering culture altogether. To meet the challenge, a cross-functional, unspecialized, experimental, adaptive and lean approach was adopted. While the traditional innovation approach at Harman International required members to be focused only on their components, in the frugal innovation project, titled Saras, members were reorganized around whole functions. The aim of Saras was to help create a cheaper product for targeting emerging market customers. The whole function-oriented strategy resulted in a holistic approach, wherein each member was able to see the whole picture at all times during the innovation process. Further, a modularized approach using a menu of predesigned features and functions ensured that features could be easily added or removed. Thus, making iterations during the innovation process became much easier. This approach ensured easy scalability, simplicity and modularity. On the other hand, the use of third-party solutions ensured that the audacious cost targets could be met.[12]

Firms must foster a culture of innovation, along with open communication and cooperation, to support frugal innovation initiatives. A firm's values must embed inclusivity and diversity of thoughts and actions, which in turn facilitate creative thinking and inclusive approaches. Open communication and increased acceptance of radical thinking leads to an innovation with greater acceptance. It also calls for more dedicated teamwork, following new ways of accomplishing routine tasks. Promoting stories of employee accomplishments are among the key drivers in frugal innovation projects. Firms often simulate the context of target users and subsequently set up incentive contests among employees to encourage frugal innovation.

The attitude of a team determines the result. Small and multidisciplinary teams with an agile mindset work better by adopting creative approaches to improve products and business models. These teams break up complex problems into modules and find solutions to each component through rapid prototyping and tight feedback loops and integrating solutions to the modules into a coherent whole. Such teams hold themselves accountable for outcomes while being committed to applying agile values, principles and practices that are required for successful frugal innovations.[13]

Employees engaged in routine tasks are unable to focus on creative efforts required for innovation. Psychological empowerment can help immensely here. Firms can institute policies and take simple measures to encourage and support innovative ideas by employees, such as maintaining an innovation box, crowdsourcing of ideas or organizing annual contests, whereby their job becomes more meaningful. Such policies help identify creative people who can thereafter be involved in innovation projects. While, traditionally, creativity has been thought of as an individual activity, the modern approach towards creativity is a peer culture. Pixar Animation Studios, presently a subsidiary of the Walt Disney Company, got good results by shunning hierarchy and promoting peer review of projects.

For frugal innovations to be embedded in a social and cultural milieu that is very different from a firm's corporate culture, suitable 'culture brokers' can be identified to bridge the gap. They 'can help build collaborative relationships among employees belonging to different departments and develop a cohesive mindset.[14]

Stakeholder-focused Approach

To ensure the success of a frugal innovation, a stakeholder-focused approach is essential. The stakeholders for a frugal innovation include the users, product developers and channel partners, the local community, environment and the industry. In 2008, Tata Chemicals launched the Swach water filter targeted at BOP families, wherein the filter was developed using rice husk, and filtration did not require electricity supply. The cost of the product was kept low, while the efficiency was high because of a precision engineering-based approach deployed by Titan, a Tata group company and a leader in the watch industry, to fine-tune the filtration process using nanotechnology. Since then, Swach has been adopted by millions of BOP families as the cost of usage is very low and the filtration process does not require electricity or running water. Tata Chemicals has partnered with many social organizations and governments to provide safe and pure water wherever it is required. For example, in 2013, Swach helped the Uttar Pradesh government's Jal Nigam provide safe and pure water to more than 100 million pilgrims attending the Maha Kumbh Mela in Allahabad.

It is necessary to involve local communities in the value chain and contribute to sustainable development by providing developing communities the increased ability to purchase products that fit their needs, reduce usage of natural resources and create inclusive economic growth. While ensuring sustainable development, socially and environmentally inclusive frugal innovations can provide profit opportunities for firms through 'for-profit appropriate technology'. It requires rethinking of entire production processes and business models, and the interventions of different stakeholders at various levels.[15]

A deep understanding of the local context is essential for ensuring the success of a stakeholder-focused approach. Some firms have set up localized R & D teams that imbibe an entrepreneurial culture and work in proximity with non-users to understand why they don't consume these products. Studies have shown that affordability and user-friendliness are particularly important for BOP customers. Such customers also prefer products that incur low or no-maintenance cost.

Knowledge capture and dissemination can play an important role here. Firms must build strong networks with partners belonging to the BOP community (e.g., supply chains or potential customers), and listen and learn. This needs to be done on a regular basis since new developments in the macro-environment can alter customer preferences, social norms and technological developments. The dialogues and discussions that employees of a firm have with locals need to be captured in a systematic manner. Knowledge management systems can help codify knowledge and store it such that it will be useful to generate insights. To foster a culture that promotes knowledge capture, job profiles of employees need to include collecting, codifying and storing knowledge that can later be useful for frugal innovations. Use of knowledge can be driven by senior leaders through meetings and discussions.

Knowledge transfer is also crucial. Knowledge gained through collaborations with various institutions needs to be passed on to channel partners with accuracy, but embeddedness can prove to be a challenge. Knowledge can remain embedded in people, tools, routines, and the networks within and outside the organization. To overcome this, it is essential that a deep understanding of the multiple knowledge reservoirs and associated sub-networks are also identified and transferred. Articulation of tacit and explicit knowledge is not easy, since individuals know more than they can explain.[16] To gain access to specialized knowledge that can help solve complex problems in the frugal innovation process, firms can partner with other firms that have complementary knowledge.

Knowledge is a socio-cultural process of learning that involves interactions with diverse people. Instead of focusing on technology, it has been recommended that knowledge needs to be garnered through social processes, whereby the meaning, purpose, values, and material aspects of users should be understood and codified. Therefore, the people involved in the process also need to be adept at socialization and building suitable relationships. By partnering with local agencies that possess in-depth knowledge of the local environment, firms can find an easier way of gaining acceptance for a radically new solution. Unfortunately, a lot of frugal innovation successes are (wrongfully) attributed to the clever use of innovative technology, but due credit needs to be given to the learning that helped design the solution.

In 2016, the Paani Foundation launched an initiative to achieve a drought-free and prosperous Maharashtra. It has already successfully touched the lives of people in 5000-odd villages across Maharashtra and helped train more than 51,000 citizens to create sustainable water storage facilities. While it has been supported by a host of philanthropic organizations, the real agents of change are the people living in these villages. The hard labour of the local people has helped transform the villages from facing severe drought to being water abundant. The change has become successful through the knowledge and training imparted by the Paani Foundation to the local villagers. Every participating village must send at least five villagers to be a part of the residential training programme. In this four-day immersive programme, the villagers are equipped with technical know-how as well as leadership skills to tackle the issue of drought. Since 2016, over 51,000 citizens have been directly trained and, in turn, have motivated lakhs more across the state of Maharashtra. For social innovations, local communities must take ownership of the problem in order to find the right solution.

It is advisable to empathize with the context of users before creating a product/solution. The goal should be to interpret the 'value' desired by customers, through dialogue between the firm and the customers. Healthcare start-up MedTech began touring rural markets in India to witness the healthcare problems faced by people and the costs that they had to bear. After visiting several villages and interacting with many customers, MedTech was able to understand the minimum performance thresholds and the features that were required and the performance dimensions, e.g., portability, that were critical considerations.[17]

To target BOP consumers in Indonesia, Unilever launched vending machines that dispense dishwashing liquid cleaner and personal care products like shampoo into any empty container that customers bring with them. Besides being a market-driven innovation, this also is an example of sensible use of energy, zero waste and active recycling, cited as one of the six principles of frugal innovation.[18] At present, these packaging-free stations are being run on a pilot basis by Unilever and a global roll-out could follow once positive impact has been established.

Frugal innovations can be transformational for developed markets as well due to the socio-economic changes and increasing constraints on resources. Frugal innovations can help ensure delivery of good public services and promotion of social and economic inclusion even in technologically advanced countries like Finland and help tackle public policy challenges in Europe.[19]

Entrepreneurial Orientation

Frugal innovations can offer customers new ways of solving existing problems and entrepreneurs will need to adopt an out-of-the-box approach to build acceptance among customers. Therefore, merely launching the innovation with fanfare and distributing the product or service through traditional channels may not serve the purpose. To begin with, it has to be acknowledged that all frugal innovations are not marketable in a big way. This is particularly true in case of innovations addressing local problems. Therefore, frugal innovations need to be designed with prudence for the purpose of commercialization. Typically, the issues affecting people, organizations, and society at large offer the best opportunities for commercialization.

Having spotted the issues, entrepreneurs need to deep dive into the causes that have resulted in these issues being overlooked.

An entrepreneurial orientation is a combination of innovativeness, proactiveness and risk taking. Innovativeness involves the generation and application of new ideas, while proactiveness is concerned with anticipation of future trends. Risk taking is the entrepreneur's willingness to commit resources on new ideas.[20] Astute entrepreneurs apply strategic management to evaluate the opportunity with respect to prevalent conditions in the market and thereafter plan to get ahead of competitors through the use of unique resources, patented technologies, efficient processes, innovative business models, etc.[21] The uncertainties surrounding a problem that needs a frugal innovation are manifold. On the one hand, the solution requires an innovative approach since carefully chosen resources will have to be deployed within cost constraints. On the other hand, the frugal innovation will need to be socially embedded so that there is widespread acceptance among user groups. These challenges call for innovativeness, proactiveness and risk taking. Let's consider some examples.

Entrepreneurs can make use of design thinking in place of traditional approaches to de-risk project outcomes. Design thinking, which involves immersion in the user's experience and segregation of voluminous data into relevant themes, helps build new insights and overcome human biases.[22] Solutions thus created are likely to be acceptable to all stakeholders. This is a crucial step towards successful commercialization of frugal innovations. Innovative solutions can also be found by entrepreneurs by making use of various findings that have not yet been utilized commercially.[23] In this regard, it is vital to establish links with research laboratories and research-intensive universities. Therefore, an entrepreneurial approach for success of frugal innovation involves exploiting information, technologies and know-how by interpreting and making use of fragmented knowledge and utilizing the knowledge possessed by specialists. Entrepreneurial orientation has helped Kolkata-based EzeRx health tech, a firm specializing in med-tech and biotechnology, to develop a device that can help diagnose anaemia in patients for a fraction of the cost compared with traditional diagnostic procedures. The diagnostic tool is called AJO and can detect anaemia and predict liver and lung problems in less than 5 seconds without extracting a single drop of blood. The process costs lesser than Rs 1. Anaemia is one of the most common global health problems affecting both developed and developing countries; it has far-reaching and severe adverse effects on human health and strongly affects socioeconomic development. Approximately two billion people, nearly 30 per cent of the world population, are estimated to be currently anaemic. Women of reproductive age, pregnant or breastfeeding women, and children are the most vulnerable populations. The device holds great promise for ensuring better health of BOP customers, and various social organizations are showing keen interest in partnering with EzeRx. The AJO generated report can be instantaneously transmitted to a medical expert through e-mail, text messaging or mobile apps. Further, the device requires virtually no maintenance and can be operated by workers after minimal training.

Vocational skill trainings, e.g., welding, spray painting, etc., are essential for providing trained manpower to a wide variety of firms engaged in manufacturing and servicing sectors. Traditional methods involving an instructor-led approach, is often found to be ineffective, especially for precision-oriented activities. A viable solution has been offered by Skillveri, which uses extended reality (XR) to develop multi-skill simulation platforms, whereby workers can get trained through machine-directed steps. The trained workers are able to work in a simulated environment almost similar to the real-world manufacturing set-up. The XR-based training has been effective in providing jobs to 50,000 trained workers at a fraction of the cost of traditional training methods, while ensuring enhanced skills that fulfil the expectations of firms in different sectors.

A lean start-up structure can minimize the risks associated with a new venture. This approach works on the principle of 'build–measure–learn', wherein a minimally viable product is created and tested on potential customers. Lean start-ups enable rapid iteration and quick learning, so that informed decisions can be made on the commercial rollout of the product.[24] Even large firms like Google and GE function by creating small teams of highly committed people for focusing on radical innovations with meagre resources. Firms build cross-functional teams and hold contests to encourage new ideas that can help improve commercialization of an innovation.

So, a right combination of innovative technology that is frugal and a business approach that is novel can lead to success. Tata Power's solar microgrid combines innovativeness, a proactive approach and risk taking. Large sections of BOP families typically living in rural areas are deprived of access to electrical power for major parts of the day. Villages do not receive uninterrupted supply of electricity from traditional sources, which are heavily dependent on coal-based power plants. To find a viable solution, Tata Power Solar has partnered with various governments and charitable organizations like Rockefeller Foundation, to offer 10,000 microgrids that would fulfil the electricity needs of more than 2.5 crore people. The venture aims to support 1,00,000 rural enterprises, create 10,000 new green jobs, and provide irrigation for over 4,00,000 local farmers.

Crossing the Barriers

Frugal approach riding on entrepreneurial orientation provides a competitive edge to organizations that are then better able to address the various challenges due to resource constraints. For instance, firms can leverage frugality-based advantages with regard to input, income and infrastructure. Input advantage involves elimination of costly inputs or processes, substituting with cheaper materials and innovating to develop a lower cost structure. Income advantage is concerned with overcoming the constraints faced by customers. While the product may be cheaper than other competing products, low incomes of customers may still prove to be a hurdle for producers. The problem is compounded by lack of access to loans from banks and other financial institutions. Though governments around the world have been taking steps to enhance financial inclusion, a large part of the population remains unbanked.

Several examples can be cited to explain how firms have leveraged frugality-based advantage. The challenge lies in first accepting the conditions under which customers (or users) are operating, and then deriving a solution that makes the best use of the existing resources (or the lack of resources). By leveraging technology that is readily available at a low (or no) cost, firms can create an advantage. A majority of BOP customers own mobile phones, which can be a point of advantage for firms promoting technologies that make use of mobile phones. Firms like Thomson Reuters Market Light and Nokia Life Services send farmers real-time information on crop prices on their mobile phones. Digital Green, started by Microsoft India, sends out videos on innovative farming practices to groups of farmers. Since the credibility of these videos is very high, the adoption rate of these farming practices is also fairly high. To create frugality-based advantage, it is essential to work around the constraints instead of making huge investments to bridge the gap in resources. The fragmented distribution systems in emerging markets present this point well. ColaLife is a socially oriented organization that leverages Coca-Cola's distribution channels to reach out to remote areas. An Aidpod kit fits between Coke bottles in crates and carries medical products like oral rehydration salts, vitamins, water purification tablets, etc. The kits can be tracked using SMS by customers, and this enables easy access for wide and scattered BOP communities living in the countryside.[25]

In order to create market-driven innovations, it is important that firms put in place systems and processes to sense market trends and align underlying technologies to serve emerging needs. When innovations strike the right balance between technology push and market pull, miracles can happen. Therefore, while planning for favourable outcomes from an innovation project, there has to be fine interplay between stakeholders' needs and market realities. In order to fulfil an unmet need, an innovative product supplemented with a service component can be a potent combination. The role of an entrepreneur is, thus, crucial in enabling a market-driven approach while maintaining the focus on stakeholders.

Going ahead, the role of networks and specialized agencies will become crucial in ensuring that innovations are market-aligned. For example, in order to promote drug discovery for diseases like tuberculosis, a market-driven and value-based advance commitment (MVAC) model is being adopted. In this system, the government, private innovators and development banks like Asian Development Bank have joined hands to make advance commitments on purchase of medicines to investors in medical research whereby the risk is mitigated. Further, this model helps ensure that the research is oriented to the needs of the market.[26] Further, social entrepreneurs can avail the support of organizations like the Schwab Foundation for Social Entrepreneurship, a community of social change leaders providing capacity building for entrepreneurs undertaking social and environmentally beneficial innovations. During the past twenty years, Schwab Foundation for Social Entrepreneurship have distributed $6.7 billion (about Rs 50,000 crore) in loans or value of products and services, improving lives of more than 6.2 crore people in 190 countries, and India is the top-most country in the list. In the future, frugal innovations that are market driven and socially embedded will surely stand a great chance of making a positive change in the lives of people.

Covid-19 pandemic has highlighted the need to redesign scientific investigations and repurpose technologies for innovative uses to find quick and affordable solutions within limited resources. From the marketing viewpoint, rapid innovation helps the early bird, more so when it addresses a problem as catastrophic as a pandemic. Almost all S&T institutions in the world started working on a war footing to control this Medusa through innovative solutions which are rapid, effective, scientifically reliable and affordable.

Some of these, like Feluda—a rapid SARS-2 virus detection technology using clustered regularly interspaced short palindromic repeats (CRISPR) Cas-9 technology, developed by Council Of Scientific And Industrial Research–Institute Of Genomics And Integrative Biology (CSIR–IGIB),[27] caught public attention at large not only because of its quirky acronym, which is based on a fictitious character in Satyajit Ray's detective novels, but also because it was rapid and cheaper at just Rs 500 per test and perceived as based on sound scientific knowledge that could even differentiate between the variants.[28] The Drugs Controller General of India (DCGI) approved its commercial launch in September 2020, giving an advantage to Tata Group, the co-developer of the technology, for early marketing. Whether the innovation becomes successful commercially is yet to be seen.

7

National Policies and Role of International Organizations

Governments will always play a huge part in solving big problems. They set public policy and are uniquely able to provide the resources to make sure solutions reach everyone who needs them.

—Bill Gates

Science, technology and innovation are fundamental to the prosperity, security and well-being of a nation, driving it on the path of all-round progress. Hence, a country's science, technology and innovation policy mirrors its priorities, social commitments and development goals. Such policies undergo changes with time in view of the changing needs, scientific advancements and international standpoints. As a country moves up the ladder of development, it reorganizes its priorities to reach the next stage and also align with other national and international agenda, which it is a part of.

National Policies on Science, Technology and Innovation

In independent India, the government accorded high importance to scientific research and technology development right from the first Five Year Plan. Separate funds were allocated to scientific research organizations like ICMR, ICAR, CSIR and ISRO; and to institutions of excellence in teaching and research like IITs, Regional Engineering Colleges (RECs, now called National Institutes of Technology or NIT), Bhabha Atomic Research Centre (BARC) and premier medical institutions like AIIMS that were established to provide a conducive environment for upstream research and building competence in different scientific fields. This paid rich dividends in the following years, in almost every sphere, be it engineering, agriculture, healthcare, space and nuclear power programmes or information technology.

The focus of India's first Science Policy Resolution (SPR), 1958, was on building a sound scientific ecosystem and it expected the flow of technology from established scientific organizations for the welfare of society. Though S&T remain important contributors to growth and development, the thrust and policy paradigm have undergone several changes since then.

In 1983, the Technology Policy Statement (TPS), 1983, was introduced, which emphasized the need to attain technological competence and self-reliance. This gave a boost to scientific development, particularly in the fields of information technology, electronics and biotechnology.

Twenty years later, the Science and Technology Policy (STP), 2003, was brought in. Having created a sound S&T base in the last millennium, this policy emphasized the need for multidisciplinary programmes and increasing investment in scientific R & D, which was more than 0.7 per cent of

GDP. STP also had features to attract private investment and highlighted the need for clear IPR regime. It made several provisions to encourage the Indian diaspora to return to Indian S&T institutions.

Having realized the importance of S&T-led innovation as the key growth driver, India declared 2010–2020 as the 'Decade of Innovation'[1] and brought its most comprehensive policy in 2013, the Science, Technology and Innovation Policy (STIP), which integrated science, technology and innovation. It recognized that scientific research generates knowledge and by providing solutions innovation converts knowledge into wealth or value. It aimed at building a strong innovation ecosystem, attracted the private sector and encouraged India's participation in all major global scientific initiatives, be it Large Hadron Collider (LHC) at the CERN laboratory or the rice genome project.

This policy identified innovations as S&T-based solutions that could be successfully deployed to build a strong economy. It was only after 2013 that concerted efforts were made to promote innovations in every walk of life and the interlinkages between science, technology and innovation were established. STIP reiterated the need for increasing R & D funding and proposed that India increase its budget allocation for S&T to 2 per cent of the GDP.[2] It encouraged private investment in S&T, which has grown to cover over 40 per cent of all S&T funding at present.[3] However, considering corporate funding in other developed countries, in the US, the corporate share in scientific research is about 70 per cent, there is much scope to increase this by bringing in policy changes and incentives. Several other policy documents, such as TIFAC Technology Vision 2020 and 2035; National IPR Policy, have also impacted the science, technology and innovation scenario from time to time and shaped up frugal innovation in particular.

The National Education Policy (NEP), 2020, envisages the establishment of a National Research Foundation (NRF) with an allocation of Rs 50,000 crore over five years to strengthen the research ecosystem in the country. Also, a draft Scientific Social Responsibility (SSR) Policy, 2019, is underway, which proposes to build synergies among all stakeholders in our scientific knowledge community and develop linkages between science and society. The SSR has the potential to enable innovative solutions to societal problems, especially of marginalized sections of society, thereby transforming the country.

The impact of government policies can obviously be seen on the prioritization of programmes and investment on science, technology and innovation. Consider the problem of vehicular air pollution in India. The country is home to thirteen of the top twenty and thirty-three of the top 100 most polluted cities in the world. The capital city Delhi tops the level of air pollution. Automobiles are the primary source of air pollution in India's major cities. As per an earlier report, the transportation sector consumes about 17 per cent of the total energy and is responsible for 60 per cent production of the greenhouse gases from various activities.[4] It emits an estimated 261 tonnes of carbon dioxide, of which 94.5 per cent is contributed by road transport. There is an urgent need to address emissions from all types of motor vehicles through creation of awareness, stricter norms and introduction of vehicles that run on non-fossil fuel. The government announced the National Electric Mobility Mission Plan, 2020, and Faster Adoption and Manufacturing of (Hybrid and) Electric Vehicles (FAME) scheme in India, with plans to have 60 lakh electric and hybrid vehicles on the roads by 2020.[5] The government is also proposing several incentives to introduce electrical vehicles (EV), particularly for public transport. This policy initiative has prompted many frugal innovations in a range of technologies from EV to hydrogen fuel cell buses, developed by Tata Motors Research Centre in cooperation with ISRO and DSIR support, by corporate firms as well as start-ups such as Log9 Materials.

Similarly, a surge in many promising innovations for managing one of the worst challenges faced by mankind, Covid-19, prompted governments to invest more in scientific R & D. The Government of India, for instance, funded 208 R & D projects worth Rs 120 crore in eighteen IITs and other S&T institutions dedicated to fighting the pandemic. Over the last fifteen years, the governments have taken concrete measures to institutionalize inclusive innovation by establishing the National Innovation Fund (NIF) in 2000, National Innovation Council (NIC) in 2010 and the India Inclusive Innovation Fund (IIIF) in 2014, and financial support through many schemes under the Ministry of Science and Technology with the aim of encouraging innovation for inclusive development. These are intended to provide solutions to the economically weaker sections of the society on health, education, food, nutrition, agriculture, energy, financial inclusion, environment, etc. Or in other words, there is a clear indication to promote frugal innovations in every sphere.

To build and strengthen an institutional mechanism for a robust and evidence-driven science, technology and innovation policy system in India, the DST's Policy Research Cell (DST–PRC) established DST–Centres for Policy Research (DST–CPR) in various academic institutions such as IISc, Bengaluru, IIT Delhi, IIT Bombay, Punjab University Chandigarh, Entrepreneurship Development Institute of India in Gandhinagar, Babasaheb Bhimrao Ambedkar University in Lucknow and many others. These centres are engaged in targeted research in a number of key areas and training of scholars in science, technology and innovation policy. There are many examples of Government of India's policy to support frugal innovation through funding and public–private partnership, such as those in the agriculture sector.

Frugal innovation, in spite of its various limitations with respect to scaling up and commercialization, has opened up a new global trend in R & D, particularly in developing countries and brought significant changes in governmental and intergovernmental policies. Frugal innovations may not always involve local innovators in development but often the success depends on building local capabilities. Therefore, combining inclusive innovation approaches with other interventions, such as public procurement, extensive training and shared use of resources becomes part of a sustainable development plan.

Frugal Innovation in International Scenario

The value of frugal innovation is recognized primarily in the context of affordability and role in sustainable development. Prof. Mark Schultz, director, Centre for the Protection of Intellectual Property, Antonin Scalia Law School, George Mason University, Virginia, US, classified innovation into three categories, putting frugal innovation in a separate category, as innovation that involves creating greater social value while stripping back the use of scarce resources.[6] He recognized that such innovations take place in resource-constrained environments in response to the needs of low- and middle-income communities, defined as the 'BOP' by management guru C.K. Prahalad.

So it is not surprising that frugal innovation, which combines affordability with quality, is at the top of every global development agenda. Varying levels of recession, experienced by most countries, concern for environment protection, and a global agenda for holistic and sustainable development have led not just developing but also developed countries to encourage frugal innovation. Recognizing the close linkages between sustainable development and frugal innovation, the United Nations (UN) and its organizations began including discussions on frugal innovation at all fora related to deliberations on sustainable development.

The global economic crisis since 2011, which is continuing unevenly across continents, has compelled industries to invest in innovation.[7] Shrinking resources forced companies to rethink their innovation strategies.[8]

As frugal and other inclusive innovations help in reducing costs for running new technology businesses, enterprises from developed countries have also started designing products for the economically weaker sections of society which constitute the biggest market segment.[9] Even then, frugal innovations remain most relevant in the context of the developing economies.

The growing impact of such products and technologies has opened up a new set of market dynamics. Viewed as a possible solution to the adverse impact of the global economic crisis on long-term competitiveness of domestic firms in developed countries, enterprises are now looking at frugal innovation to develop products and services for a new category of consumers in new territories. In fact, the relatively stable and better economic performances of China, India and Indonesia vis-à-vis some of the slower-growing Organisation for Economic Co-operation and Development (OECD) countries were attributed to the adoption of inclusive innovative technologies and methods.

While innovation has always contributed to a nation's economic development, in the past it has mostly been classical innovation. But having felt the impact of global recession and climate change, the international community has realized that innovations through conventional means have limited reach and access to deal with the multiple challenges of inequality, scarce resources, migration, sustainable development, etc. Furthermore, many emerging economies, including Brazil, India and Indonesia, have demonstrated that frugal innovations not only support sustainable development but also add to the economic well-being of a country. It has been observed[10] that international centres of excellence, which have research interests in developing and emerging countries and support technological innovations, can accrue benefits for both their own countries as well as their international partners from developing countries through research spill overs and business opportunities. Supportive policies, partnerships and dedicated research funding can create a win-win situation.

One of the drivers for innovation in developing countries has been the desire to catch up with technologically developed countries and thereby trying to develop a variety of innovation capabilities across sectors through existing institutional systems. Since innovations from developed countries relied on high-quality infrastructure and skilled workforce, and often were not designed for conditions in developing economies, their adoption added more challenges. Technologies not aimed at fulfilling the needs and means of the economically weaker sections of the population only widened the inequality.

Recognizing the value of frugal and other inclusive innovations in sustainable development, the international community is embedding frugal innovations within the framework of international cooperation. The UN has defined frugal innovation as 'a distinctive approach to innovation, which minimizes the use of resources in the development and delivery of innovative products, thus resulting in low-cost innovation that can become drivers of growth, especially in developing countries'.

While conversations in the international arena continue to focus more on sharing of best practices and capacity building, institutionalizing frugal and similar innovations in sustainable development policies is gaining international traction and influence too. The key conversation on innovation is continuing at the UN and its various organizations. Of these, UNCTAD is the nodal institution that deals with both policy and implementation of recommendations of member states.

Against the backdrop of increasing concerns over climate change, which resulted in the historic adoption of the 2030 SDGs by 193 member states in September 2015, discussions on innovation have become an integral part of the SDGs. Two key points that emerged from these discussions are the identification of innovations that are likely to help accelerate attainment of SDGs, and the need for cooperation and collaboration between governments, international agencies and the private sector for delivery of services through such innovations. The characteristics of frugal innovation that appeal to the global community are:

- affordability, even to the economically weaker sections, by stripping all non-essential features
- involvement of stakeholders at the grassroots
- being inherently resource conserving, it contributes to a greener environment

In this context the World Trade Organization (WTO) observed that 'the focus is on making better things, not just cheaper things or not just downgrading existing innovations: rather remodelling goods and services, which is not just low cost, but also high tech'.[11]

At the UN General Assembly (UNGA), the UN's highest decision-making body, member states (mostly from developing countries) urged the international institutions to take cognisance of frugal and other inclusive innovations which benefit societal groups that have previously been ignored. International institutions have, in turn, been urged by the UNGA to adapt policies on science, technology and innovation based on the new technological trends and developments, with the UN continuing its role as a facilitator.[12]

Frugal innovation has opened up a new global niche in R & D. The constraints on conventional R & D in developing countries due to various factors have led to a surge in frugal innovations and the international community has acknowledged that such innovations are helping reduce costs for opening and running new technology businesses in spite of many challenges, such as region-specificities of technologies and difficulties in scaling-up. Possible solutions include use and support through digital technology, microfinancing and PPPs in research, technology development and commercialization. In fact, a 2017 UNCTAD study has underlined this, recommending efforts to effect changes in policy in investment as well as to overcome resource constraints.[13] Acknowledging the role of the private sector, it pointed out that the private sector in developing countries is mostly made up of small and micro enterprises, and the informal sector is larger in these economies than in the developed ones. Developing countries, therefore, need to strengthen their own innovation systems or rely on access, transfer and absorption of foreign knowledge and technology that may come at some political and economic costs.

Despite the ongoing recession, global spending on innovation is still resilient. At the same time, developing countries need to factor in the twin challenges of increased protectionism and diminishing public support in developed countries for R & D in innovations. As the international community becomes increasingly aware of the benefits of such innovations, opportunities for financing proposals and projects are opening up. The United Nations Framework Convention on Climate Change (UNFCCC) technology Executive Committee, the Climate Technology Centre & Network (CTCN) and the Green Climate Fund (GCF), have already collaborated to create a protocol for receipt of proposals for research on climate technology incubators and accelerators to be managed by the board of the GCF. The GCF is already funding projects, including those for small renewable energy companies, in Africa that accelerate tech innovation.[14]

India, a leader in frugal innovations, has taken many measures to promote these and other inclusive innovations. The international community has taken note of India's various policies and schemes, including its annual Festival of Innovation, started in 2015, which celebrates creativity and innovation at the grassroots. Apart from various organizations supporting the festival, United Nations Educational, Scientific and Cultural Organization (UNESCO) has extended collaboration and committed to review its policy in inclusive innovation.[15] Many such governmental and non-governmental initiatives, nationally and internationally, encourage individuals and organizations to find effective solutions through frugal innovation. BRAC, Bangladesh, considered one of the world's largest NGOs, arranges the Frugal Innovation Forum (FIF) every year with a new theme and has been doing so since 2013. It provides a platform for leaders, development practitioners, funders and innovators from the Global South to connect and explore solutions to the world's toughest challenges. India's central role in pioneering frugal innovations has been presented in many such international conversations.

Many countries support frugal innovations through recognitions and awards, where market opportunities are lacking. The Longitude Prize for innovation in the UK is one such example, which is awarded for developing an affordable, accurate and fast point of care test for bacterial infection that is easy to use anywhere in the world. Advance market commitments also encourage a cascade of multiple manufacturers, which ensure both competition and sustainable production. For example, for the Pneumococcal Advance Market Commitment of 2009, donors committed funds to guarantee the price of the vaccines. This provided an incentive to vaccine manufacturers to invest in research in return for providing vaccines at affordable prices.

The concerted efforts of the Indian government towards creating a conducive ecosystem to encourage frugal innovation through various schemes and incentives are cited and deliberated in other countries to facilitate policies for supporting frugal innovation[16] and its model of promoting frugal innovation are much appreciated and even adopted in other developing economies.

Spread of Frugal Innovations

More and more countries are recognizing frugal innovations as sustainable options for economic growth and development. China has embraced the mantra of frugal innovations in a big way to become a global leader in technology, which has provided a strong foundation to its economy.[17] Focus on innovation in China's energy system led it to not only report 100 per cent rate of grid connectivity, but also an increase in the proportion of renewables in total electricity generation from 16.2 per cent in 2005 to 23.4 per cent in 2015. Colombia used its innovation policy to overcome inequality and contribute to the peace process (SDGs 10 and 16), whereas focus on healthcare innovation in Cuba enabled it to develop the world's first human vaccine against Haemophilus influenzae type B to contain a synthetic antigen.

Given the health sector's pivotal role in the socio-economic well-being, it is hardly surprising that this is where the largest number of frugal innovations are to be found.[18]

The WHO itself utilizes some of these new and reliable low-cost medical devices and services, especially adapted to the needs of developing countries. The main challenge in providing good healthcare is access, as economically weaker sections simply cannot afford the cost of medicines or some of the services required, despite welfare measures such as free treatment or subsidies offered to them by the government. In other cases, imports from developed countries are often unsuitable for developing countries. For example, gadgets consuming high power meant for developed countries are unsuitable in developing countries where power supply is often erratic or not available at all. The gadgets may remain unused, or not function in hot, humid and dusty conditions prevalent in many developing countries, as noted in WHO's Guidelines for Health Care Equipment Donations.

The Indian pharmaceutical sector, which supplies over 50 per cent of the global demand for various vaccines, 40 per cent of generic demand in the US and 25 per cent of all medicines in the UK is an important player in the global pharmaceutical industry. India has also made good progress in healthcare innovations with a large number of need-based, high efficiency and affordable medical gadgets, devices and diagnostics (see chapters 2 and 3 for details), which have good domestic as well as export market potential. This is due to investment and policy support in the form of export incentives, relaxed taxation rules and research funding from various DST/DBT schemes. However, similar progress could not be made in the case of sophisticated, expensive and heavy-duty medical equipment, which is still manufactured by a handful of manufacturers based in developed countries.

The appeal of frugal innovation in the healthcare sector in developing countries is not just restricted to affordability but also safety and ruggedness. In 2013, the coordinator of the medical devices equipment in WHO emphasized that 'a surgeon who needs to use an anaesthesia machine can use one that is simple, but it must be safe and effective, apart from being affordable and appropriate for the local hospital setting'. In addition to being more reliable, the medical equipment should also be more accessible. The WHO continues to encourage frugal innovation in the development of such equipment in cooperation and collaboration with its member countries.[19]

Demand for a large number of affordable but more importantly reliable ventilators at the time of the Covid-19 pandemic and the surge of frugal innovations that followed within two to three months was an affirmation of the national scientific capabilities and the importance and relevance of frugal innovations across global economies. This also highlighted India's opportunities to become a world leader in pharma and healthcare sectors with the right policy support.

Transnational giants like GE consider local conditions while developing new products. For instance, one of the most successful frugal products from GE is the Lullaby Warmer, which provides direct heat in an open cradle and is used to help newborns adjust to room temperature. The original product available in the US cost about $12,000 (approximately Rs 8.76 lakh) but the modified Lullaby Warmer for India was deliberately made at a lower cost and had additional features such as monitoring an infant's pulse and weight. The modified Lullaby Warmer which was launched in India was priced at less than Rs 1,00,000 and introduced as 'designed in India, for India, that can also benefit the world'.[20] Because of sound technology at affordable cost, it is sold in many developing as well as developed countries.

Digital media has played a key role in accelerating the reach and spread of frugal and other inclusive innovations, and acceptance by the international community as key transformers for sustainable development. The simple advantage of frugal innovation is that it directly addresses the welfare of lower income groups. While similar benefits are also obtained through grassroot innovations, the methods by which the benefits reach the economically marginalized are different. Consumers are the beneficiaries in frugal innovation, while the producers are the beneficiaries in grassroot innovations. Above all, frugal innovation has served to underline that the more innovations are democratized the more benefits accrue from them.

An alarming rise in income inequality over the last three decades in many OECD countries has put the focus on social and industrial inclusiveness through concrete policies on innovation and inclusive growth. The OECD supports a number of projects examining how innovation can serve inclusive development in different economies.[21] Such projects have been able to successfully draw the attention of the international community to economic disparities in various sectors and establish that without inclusive growth, a country's economic and social well-being are adversely impacted. Conditions that accelerate democratization include creation of platforms, reduction of barriers on IP policies and promotion of small enterprises, formation of like-minded groups undertaking similar frugal innovations, encouraging skill sharing between different innovation groups, accessing S&T from the public sector and improved infrastructure.

A New Model is Emerging

While promoting grassroot and frugal innovations, there is scope for combining the classical innovation approach with the new pro-sustainable approach by creating a 'hybrid pathway to sustainability'.[22] A few interesting case studies and their impact have been reproduced from the 2017 UNCTAD study.

The United States Agency for International Development (USAID) started a programme to counter the outbreak of the dreaded Ebola virus disease in 2014 in West Africa, called 'Fighting Ebola, A Grand Challenge for Development' that reached out to traditional as well as conventional health specialists for ideas on treating and controlling the disease. Out of 1500 ideas, fourteen were identified in areas of protection of healthcare workers, comfort and care of patients, improving healthcare workers' tools, decontaminants, etc., including a wearable patient sensor used in Sierra Leone in 2015. Similarly, a low-cost battery-infusion monitor developed by John Hopkins University was used in 2016 in Liberia and Guinea encouraging DuPont to mass produce it. The recent outbreak of coronavirus (Covid-19) has triggered a similar urgency in the global scientific R & D institutions both in the public and private domain to come up with effective, affordable and quick solutions to contain this killer disease threatening half of the world's population.

The United Nations Entity for Gender Equality and the Empowerment of Women has developed a programme called 'Buy from Women', a mobile-enabled supply chain enterprise platform for cooperatives, connecting women to information, finance and markets. Similarly, India's e-Choupal connects agriculturists to their supply, delivery and value addition chain, providing information on weather, market prices, etc., via SMS. There are several examples of the Indian model of frugal innovation which are proving effective in other countries.

The Barefoot College in Tilonia, Rajasthan, was founded by the social entrepreneur Sanjit 'Bunker' Roy, who has been championing a bottom-up approach to education and empowering the rural poor since 1972. He is engaged in capacity building of rural people for sustainable development. The training of African solar grandmothers by their Indian counterparts is one such example of international cooperation in adopting frugal innovation for sustainable development. In this simple model, women from around the world are recruited and given hands-on training to instal and maintain solar lighting and power in their home villages, thanks to Roy.[23]

The Barefoot College has now trained batches of rural women from twenty-four countries with little or no education. These women with a hunger to learn are trained for six months using simple innovative teaching medium and technologies in how to assemble, instal and maintain small photovoltaic (PV) systems that could bring solar electricity to their houses and community. The barriers of language and lack of formal education are addressed through colour coding, gestures, and demonstrations. This innovative technology transfer has successfully electrified 35,000 homes in the Middle East, India, Asia, South America, and Africa.

An example of a contemporary grassroot innovation movement in developed countries, post-recession, is the maker movement. It is now a global driver of informal experimentation with technologies such as software, microelectronics, robotics and digital fabrication. This movement links traditional knowledge about carpentry, metallurgy and mechanics with new skills and technologies, such as software programming and basic electronics. 'Hackerspace', 'FabLab' and 'MakerSpace' are terms frequently used interchangeably to explain community-operated and primarily non-profit workspaces where people with common interests in innovation can meet to cooperate and collaborate. In this context, it is interesting to note the evolution of FabLabs in 2001 at MIT's Centre for Bits and Atoms, wherein computers and related tools are provided to allow the community to undertake innovations. There are specific terms and conditions to follow, as the FabLabs are under the MIT franchise. According to Fab Foundation, there are approximately 2000 FabLabs worldwide till date, including in developing countries.

Another key example of frugal innovation is the One Million Cisterns project, created by a network of civil society institutions, to provide a significant number of water cisterns in a large semi-arid region in north-eastern Brazil. The cisterns receive and store seasonal rainwater for both personal and productive needs (such as agriculture) throughout the arid season. This project is the outcome of collaboration among more than 700 institutions, social movements, NGOs and farmers' groups, coalescing into the Semi-Arid Association along with Social Technologies Network, the Brazilian ministries of environment and social development. This project was built by farmers and labourers, and it empowered local communities, providing autonomy from local governments and water suppliers. Since its inception in 2003, 5,90,000 water cisterns have been built by locals.[24]

These innovations, while well-known worldwide, have become established best practices through the platform of the UN institutions where they have been discussed, reported upon and supported through the UN's programme of technical cooperation, and replicated in other countries. In fact, through international institutions, inclusive innovations have found international recognition. While the examples mentioned above have sown the seeds for the international community to strengthen discussions on frugal and other inclusive innovations, including efforts to institutionalize cooperation in inclusive innovations, care is needed to ensure that established institutions avoid imposing their objectives as all such innovations thrive due to independence from bureaucratic procedures and institutional mechanisms.

Social innovations have received substantial academic and policy attention as these are meant to deal with social needs and improving human well-being and have beneficial impact on frugal innovation efforts. Over the last decade, national and international policies to encourage, promote and nurture social innovations have been formulated. The sectors of focus are youth development, employment, capacity building and education. For example, in the US, the Social Innovation Fund offers grants and technical support to innovative initiatives to replace ineffective public programmes. The European Commission, through its Employment and Social Innovation Programme, offers networking support, an annual innovation competition and related research funding. Like sustainable innovations, social innovations also have challenges, notably that of economies of scale. There is also the danger of adverse impact of credibility of innovators in certain cases.

Time banks, which are systems of exchange of reciprocal service, are a new method of fundraising and occupy a space between the state, private sector and civil society. Through a time bank, community members support each other through online platforms. Members can deposit time credits in a time bank and withdraw equivalent support later. They have enabled members to overcome loneliness and improve health and well-being. Today, organized time banking takes place across 1000 banks in more than thirty countries, including China, Russia, various countries in Africa, Europe, North America and South America. A well-known time bank in the UK is the Give & Take Care, which works with a charity organization for senior citizens to match the skills of the caregivers with the needs of the elderly. Every hour of volunteer work may be logged in the time bank and exchanged for care in later life.

Notwithstanding the encouraging trend in funding of innovation proposals and projects, funding (both national and international) for SDGs remains insufficient till date. While there has been global attention paid to linkages between governments, universities and industry in innovation policies, nurturing linkages with investors have been inadequate, especially in developing countries. This is a challenge that needs addressing by the international community.

In terms of establishing regulatory frameworks, mechanisms on public procurement can act as an incentive to support innovations directed towards sustainable development. Similarly, IP rights can be a powerful tool to stimulate innovation in both developed and developing countries, provided frugal innovations are given some incentives. As frugal innovations are mainly taken up by small enterprises with limited resources, such support will be needed.

The World Intellectual Property Organization (WIPO), which can directly impact policies, has recognized the positive contribution of frugal innovation in the economy and empowerment of the marginalized population, and is providing incentives for innovators. In an interview in 2017 to the *Economic Times*, Francis Gurry, director-general of WIPO, agreed that there has been a shift in innovation over the last twenty years from domination by North America and Europe towards Asia in terms of magnitude. Japan, China, Korea and India are central to this.[25] Replying to the comment that one of the issues in innovation is that it does not address the problems of the poor or the developing world, Gurry drew attention to the fact that a lot of innovation is science-based or magnitude-based, and producer driven. He said that at WIPO there are several programmes that provide incentives to developing countries on patenting, such as machine-based translations (which are less costly) and free legal aid.

In the Indian context, the IP system is only one of the many factors affecting innovation. The others are human resources, education system, finance, infrastructure, governance, etc. Apart from not being able to bear the costs of protecting their

IP, the stakeholders of inclusive innovations also cannot afford and do not have the capacity to litigate infringement of their IP. As in the past, India's successful efforts to revoke the patent on turmeric issued by the United States Patent and Trademark Office (USPTO) had faced and overcome similar challenges in establishing that turmeric, a native Indian plant, had been used for centuries by its people for healing wounds.[26]

The present IP system may be more suited for developed countries than addressing the interests of developing countries. Therefore, success of frugal innovation will depend more on the capacity of the innovators to cut costs, rather than on protection of IP, which does not necessarily guarantee technology transfer. Besides, frugal innovations, often built upon existing technology, can seldom be patented. The National IPR Policy of India rightly aligns to promote creativity and innovation.

Frugal innovation could open up a new growth regime, with its own norms through the production of cheaper goods which correspond to the needs of the poor in both developing and developed economies. As the consumption trend that triggered a growth regime after the Second World War in developed countries has now reached a plateau and in many cases reduced, frugal innovation can be considered as a possible alternative to boost consumption and push demand for sustainable growth.[27]

India's Role in Frugal Innovation

The international community continues to turn to India for successful examples of frugal innovation, in both products and services.

India's role in guiding discussions in international fora is significant not only for its economic development, but also for creating a global road map for frugal innovation. The Global Innovation Index (GII) releasing its figures in 2021, showed

India moving steadily up the innovation indicators. India, with a rank of eighty-one in 2015, moved up to sixty-six in 2016, fifty-two in 2019, forty-eight in 2020 and is now ranked forty-six (as on 2021). India is in the top position in Central and Southern Asia, with improvements having been affected in institutions, human capital and research and market indicators. India's position has, according to the report, outperformed on innovation relative to its GDP per capita for nine years in a row. At the same time, challenges need to be overcome to improve poor logistics, increase the employment of qualified women and improve the qualitative performance of print and other information media.[28]

As an increasing number of countries are turning to frugal innovations for economic solutions and sustainable development, it is clear that no country can sustain such innovation in isolation. Therefore, both developed and developing countries, regional organizations and transnational enterprises are turning to the UN and its organizations for international and institutional support. Side events and discussions showcasing frugal innovations are now held regularly during meetings of the UNGA, UNCTAD, UN Commission on Science and Technology for Development (CSTD) and the like.

The willingness to share experiences, best practices, built-up capacities and desire to be part of the policymaking process in this vital sector have resulted in greater cooperation and collaboration by member states across continents. A recent example is that of Thailand, which in 2019, at the twenty-second session of the CSTD, in collaboration with UNCTAD, organized its side event to showcase current Association of Southeast Asian Nations (ASEAN) cooperation on science, technology and innovation for sustainable development, including frugal innovation, both within and beyond the region.

The global appeal of frugal innovation is creating new opportunities for employment. In August 2017, the director-general of the International Labour Organization convened an independent Global Commission on the Future of Work[29] which highlighted that technological advances in AI, automation and robotics will create new jobs, but those who lose their jobs in this transition may be the least equipped to seize new opportunities. On the one hand, the greening of our economies will create millions of jobs as sustainable practices and clean technologies are adopted, but on the other, jobs will disappear as countries scale back their carbon- and resource-intensive industries. Changes in demographics are no less significant. Expanding youth populations in some parts of the world and ageing populations in others may place pressure on labour markets and social security systems, yet in these shifts lie new possibilities to accord care and encourage inclusive, active societies. Accordingly, the commission proposed a human-centred agenda for the future of work that strengthens the social contract by placing people and the work they do at the centre of economic and social policy and business practices. This agenda consists of three pillars of action, which in combination is expected to drive growth, equity and sustainability for present and future generations:

- Increasing investment in people's capabilities
- Increasing investment in the institutions of work and
- Increasing investment in decent and sustainable work in line with the 2030 Sustainable Development Agenda.

The world is on the path of sustainable development, embedding frugal innovation as a logical approach. As Leach (et al., 2008) had observed in the context of Global South, innovation, and S&T need to be made part of a participatory process of development, where citizens are involved, not just as recipients but also for knowledge share, technology development (and selection) and delivery.[30] Even if the pace of such development and market growth are slower than the current pace of breakneck competition, it is more likely to be sustainable, creating value for all, but most for those at the bottom of the pyramid. A country that pioneered frugal innovation, India continues to participate actively in various platforms of international cooperation for frugal innovation to protect and promote the well-being of India's frugal innovators—both producers and consumers, and towards attainment of an equitable and prosperous society through sustainable development.

The 'Mission Innovation' (MI), an S&T-led intergovernmental initiative, which was launched by twenty countries in November 2015, is an example of international collaboration to achieve the common goal of a cleaner world. India is co-leading three of the eight challenges being addressed under this Mission. 'MI initiative reflects the collective desire of nations to work together to solve shared challenges to provide affordable and reliable clean energy.'[31] Currently, there are twenty-two member countries and the European Commission in this mission. The member countries and the European Commission are committed to doubling government funding and enhancing international engagement for clean energy R & D. Two of the challenges, 'Smart Grid' and 'Off Grid Access to Electricity', are being implemented in India by the DST, whereas the 'Sustainable Biofuels' challenge is being led by the DBT. Through the platform of MI, India has taken the lead in developing a framework to promote affordable clean energy technologies with active participation of scientists, researchers, academia, industries, as well as private organizations. Sustainability and affordability (frugality) of technologies are central to these programmes.[*]

---

[*] Authors gratefully acknowledge the use of UNCTAD, 2017, report.

8

Science and Sustenance: The Path Ahead

You cannot escape the responsibility of tomorrow by evading it today.

—Abraham Lincoln

In September 2015, the UNGA adopted the 2030 agenda for sustainable development with seventeen SDGs, which were agreed upon by 193 countries.[1] Building on the principle of 'leaving no one behind', the agenda emphasizes a holistic approach to achieving sustainable development for all.

However, in spite of earnest global cooperation and multi-programmed efforts, the annual SDG report 2019 noted that even during the pre-Covid period, good progress could be made only in some areas. While there was overall reduction in extreme poverty, progress in widespread immunization, decrease in child mortality rates and increase in people's access to electricity, all of which impact one or more SDGs directly or indirectly, it noted that the global response has not been ambitious enough in other sectors, 'leaving the most vulnerable people and countries to suffer the most'. This indeed is an insightful observation, which brings forth the relevance and need for equitable development and opportunity for all. Perhaps more so in the post-Covid world.

The major problems ailing the world today, be it air and water pollution, healthcare, food and nutrition security, natural calamities resulting from climate change, education or employment, impact the well-being of the people at BOP more than those on economically safer grounds. As António Guterres, UN secretary-general, commented in the annual SDG report, 2019, 'It is abundantly clear that a much deeper, faster and more ambitious response is needed to unleash the social and economic transformation needed to achieve 2030 goals.' This is where frugal innovation is going to play a vital role, provided a conscious effort is made by all concerned.

The question is not whether frugal innovations can be transformational, that has been well established in several instances, but are we doing enough to promote frugal innovations? Though the national policy frameworks, as well as international conventions, acknowledge the need to put development of accessible and sustainable innovations on high priority, often these are not perceived as the preferred goals by the scientific community. One of the primary parameters to evaluate the performance of S&T professionals is through the worth of their publications. While India is the third largest publisher of science and engineering (technology) papers in peer-reviewed scientific journals, only behind the US and China, its citations are much below the world average. Much is needed to change the mindset of the scientific fraternity to encourage innovations that are useful to people at large, rather than those oriented towards what is recognized as 'upstream', using expensive tools and technologies, that are likely to be published in scientific journals having high impact factors. Strong industry–academia partnerships through contract research, right from the identification of the problem and ideation to technology development and validation, could be one way to promote creation of frugal technology, though it has the pitfall of being more market driven than of societal value. This, to some extent, can be safeguarded by partnering with the user/consumer forums and local small/medium entrepreneurs, who are often more conversant with users' problems at the ground level and also understand the limitations in adopting new solutions. It would, however, be relevant to reiterate once again that use of sophisticated tools and technologies could be integral to research programmes aimed at developing products for the consumers at the BOP. A balance between fundamental and applied research and enabling S&T policies can foster frugal innovations in a more systematic manner for wider reach. Presently, only 10–15 per cent of frugal innovations, such as Nokia 1100 or GE LOGIQ, capture a sizeable market share. A large percentage of these address situation-specific solutions and have seen only limited commercial gains. The remaining, even with high societal benefits, may be of little economic value to the technology provider, such as Paperfuge or Foldscope. Therefore, different business models are needed for frugal innovation, based on ecosystem advantages, to convert some of the societal and ecosystem benefits into monetary gains. Moving forward, world economies at large, not just the developing ones, will need to reorient their development and growth plans in sustainable mode. Prioritization of key areas of development and promotion of frugal innovation will be inherent to such plans.

Priorities by 2030

The world is going through an era of unprecedented technology advancement, racing at the same speed as the human imagination. Adoption of new technology is leading to rapid changes in people's aspirations and lifestyles. As a result, major restructuring of social, cultural, industrial and economic norms is expected within the next decade. Some of the technologies that are likely to be part of our daily lives by 2030 include self-driven cars, use of microgrids for household electricity supply, cryptocurrencies, preventive and personalized medicines, laboratory synthesized meat, robotic surgeries, ethical use of genetic engineering to cure non-curable genetic and degenerative disorders, 3-D bio-printed organs and perhaps many more. Though these are more likely to capture markets of developed economies, dominance of AI and big data analytics-based solutions will become common across economies in everyday life.

Adoption of cutting-edge technology, while opening new markets, can also contribute towards bigger social inequities and, hence, give rise to new challenges. Frugal innovations are expected to play an important role in mitigating these challenges. It is difficult to accurately predict the changes, but one thing is clear—frugal innovations will be more norm than novelty in finding solutions to the problems impacting people's lives in general, and the well-being and sustenance of the ecosystem in particular.

The key areas that the world today cannot afford to ignore and which call for immediate action include:

- hunger and nutrition
- climate change
- air pollution
- scarcity of safe drinking water
- human health and
- capacity building

Despite the threat, these challenges offer unique opportunities for development with frugal innovations paving the way. This is because from a consumerist point of view, not just developing economies, but the world at large will need to adopt a sustenance orientation. The current trends, national policies, polity and intergovernmental conversations on science, technology and innovations; changing aspirations, lifestyles and needs of people; and growing awareness about the value of sustainable technology and eco-friendly lifestyle are going to be the key drivers of frugal innovations by the next decade. Though all aspects covered under SDGs are crucial for global progress, from a developmental perspective, the following could be considered the priorities for frugal innovation:

Priority	Preferred Adoption
clean air	all economies
safe drinking water	developing economies
food and nutrition	developing economies
vaccination and diagnostics	all economies
affordable healthcare for all	developing economies
web-based skilling and virtual education	all economies
eco-friendly and affordable dwellings	all economies
affordable use of clean and renewable energy	all economies
affordable, eco-friendly and efficient transport system	developing economies

Science-led frugal innovations can bridge the gap between people's dreams, needs and fulfilment of goals for sustainable development. It is only possible if every nation focuses on the needs of disadvantaged persons and creates enabling infrastructure, systems and facilities so that such people can contribute and prosper in society with dignity. Enabling policies, funding, partnerships between the government and private agencies, and efficient handling of IPR, validation and prompt approval by the agencies concerned can help unleash the power of frugal innovations and establish India as a world leader. In all this, strengthening scientific research will be vital. While most European and North American countries spend between 2 per cent to 4 per cent of their GDP on scientific research and technology development, most Asian and African countries are far behind with around 0.5 per cent of their GDP being spent for the same. Realizing that investment in science helps in their economic growth, even emerging economies of the world have been steadily increasing their expenditure on scientific innovations in the past few decades. China, attaching high importance to innovation, increased its national expenditure on scientific research from 0.63 per cent of GDP in 1996 to 2.07 per cent in 2015, whereas India's expenditure has remained around 0.70 per cent for the past twenty years and needs to be raised substantially, if not doubled.[2] However, it will only be logical to create short-term plans in sectors where India already has gained a scientific lead, such as pharmaceuticals, healthcare, software and agriculture.

Affordable Healthcare and Wellness

Growing population, increase in life expectancy, excess consumerism, lifestyle maladies, disease burden, etc., are all creating tremendous pressure on the system to provide better healthcare facilities. While eradication of a number of communicable diseases, better pre- and post-natal care, and better availability of food and medicines will continue to help people live longer, affordable and accessible healthcare remains a challenge for BOP sections.

India, one of the leaders in the pharmaceutical industry, is the largest provider of generic drugs and meets more than 50 per cent of global demand for various vaccines. The CSIR institutes, being at the forefront of drugs and pharmaceutical research, have developed many new drugs, cost-effective and innovative processes for around twenty-five generic drugs and devised new diagnostic tools, besides standardizing over fifty herbal drugs.[3] Together with a strong private sector, India has tremendous opportunities to consolidate its position in the post-Covid scenario by adopting frugal approaches in innovation, processes and bulk manufacturing.

With the healthcare sector expected to grow at 20 per cent in 2020–21,[4] private investment in this sector has already shown steady growth and dominance, which was previously dominated by the public sector. But there is still a huge gap between the actual need and availability of affordable and accessible healthcare. It is estimated that by 2025 India would need investments worth 3 per cent of its GDP[5] to bridge the existing gap.[6]

The pandemic due to novel coronavirus Covid-19 brought forth, like never before, the urgency to upgrade our healthcare systems. To contain this devastating disease, universities, research establishments, firms, etc., around the world quickly got into the act of developing vaccines and other cures. India is not only leading thanks to its capacity to mass produce vaccines at an affordable cost but has also, in a short time, developed affordable diagnostic methods for rapid detection of the virus. The validated frugal solutions and other innovations developed during the Covid-19 fight shall help the country tackle similar major challenges in future.

In the wake of the crisis, the nation also found an opportunity to assess and upgrade its medical infrastructure. One of the critical facilities lacking is the availability of a sufficient number of ventilators. India is estimated to have 40,000 to 50,000 ventilator units in a country of 130 crore people. A team of innovators from IIT Roorkee in collaboration with AIIMS Rishikesh caught the attention of medical professionals and commercial manufacturers alike with news of the development of an affordable, efficient and simple-to-use portable ventilator unit. This machine can be crucial for the survival of those suffering from acute pulmonary problems, as in Covid-19 patients. Named PranaVayu, it is equipped with state-of-the-art features, such as real-time spirometry and alarms, and can be used for all age groups, especially the elderly. As it does not require compressed air for functioning, it can be especially useful in temporary facilities in emergency situations. The manufacturing cost per ventilator is estimated to be around Rs 25,000, whereas traditional ventilators cost in the range of Rs 10 to 15 lakh.[7]

Another potential area of healthcare where India can break through the market is that of telemedicine and web-supported medical consultation. While a few firms have begun offering telemedicine solutions for cheaper access to dependable advice from expert doctors on health problems, a holistic solution can evolve only with participation from civic authorities and growth of a supply chain. Mohalla Clinics in Delhi are a good example of the concept. Utility of telemedicine became evident during one of the longest nationwide lockdowns in India to contain the spread of Covid-19, particularly in view of a large proportion of the population being above sixty years, which as per Census 2011 was estimated to be over 10 crore. The digital and the social worlds need to be merged to create affordable and effective geriatric and healthcare solutions.

Self-diagnostic kits for various ailments at affordable prices can help BOP consumers get easy access to tests that otherwise burn a hole in their pockets. The departments under the Ministry of Science and Technology, through funding platforms like BIRAC, are supporting a large number of scientific teams working independently (such as Dr Dendukuri, see chapter 1 for details), in S&T institutions (such as Prof. Mehta, see chapter 2 for details) or at incubators for developing affordable, easy to operate and accurate diagnostic tools for early detection of serious diseases, leading to better control. Most such projects target ailments of higher mortality risk such as cancer, hepatitis B, etc. However, diagnostics for a large number of other common communicable diseases have equal importance and are needed in countries like India, where unsanitary shelters, and lack of sanitation and safe drinking water are primary causes of several diseases. There is, thus, a strong need for institutional efforts to promote frugal innovation in the health and wellness sector.

India has already shown its strength in innovative medical care services (e.g., Narayana Health and Aravind Eye Care), diagnostic technology (e.g., Dengue Day One Test, OncoDiscover and Rotavac), affordable prosthesis and implants (e.g., Jaipur Foot, Aum and Poorti). Devices like ToucHb, a non-invasive diagnostic procedure innovated by Mumbai-based start-up Biosense to screen for anaemia without drawing a blood sample, can have far-reaching implications in improving public health, particularly in the case of women and children. Smartphone-based diagnostic tools for common parameters such as blood and urine sugar, lipids and Hb have great scope and are expected to be in regular use soon. India, thus, has an 'opportunity to leapfrog a lot of the healthcare problems that developed nations are grappling with'. This again has assumed greater relevance in the Covid-19 pandemic situation.[8]

Provision of supportive healthcare by trained health workers and life-support systems at home could become more common than personalized medicines in India, where the population of people over sixty, living in nuclear family units, is increasing steadily. Firms like Athena Healthcare have already begun providing such services, but more is needed. Not just the technology, but innovative micro-health insurance products are also expected to play a big role in making medical care available in time, as demonstrated by the success of Yeshasvini, a health insurance scheme of the Karnataka government for cooperative farmers who get quality medical care, including cardiac surgeries, performed at reputed hospitals. We need more flexible models and necessary support from the government or corporate houses under CSR, keeping in view the financial capacity of BOP patients.

Agriculture: Producing More From Less

India is among the top two producers of several agri-commodities and the second-largest agrarian economy in the world today. Developments based on scientific and innovative technological solutions have helped transform the country from the status of 'ship-to-mouth' in the 1950s and 1960s to a food-secure nation. This has been vital in reducing poverty and hunger levels by 50 to 70 per cent (R.B. Singh, 2020).[9] However, considering that India, with only 2.3 per cent of the world's land and less than 4 per cent of the global freshwater, is required to feed nearly 17 per cent of the world's population, future growth in agriculture can come from:

- increased productivity per unit area
- reduced post-harvest losses
- minimized use of water and other resources

- improved nutrition and value addition to agricultural produce
- mitigation of climate change
- reduced use of chemicals
- precision farming

And this will have to be achieved without causing more damage to the agro-ecological system or accumulating carbon footprints.

Recognizing the importance of frugal innovations, in an open letter to the UN, Group of 20 and national governments, the leaders in global agriculture research and policy, acknowledged that 'the revolution in ICT and in biology can help reimagine the food and agricultural systems to provide food security to the poor, and to transform the sector by reducing its environmental and climate footprints. Disruptive innovations are needed to increase productivity and income through precision farming and timely delivery of inputs to farmers' fields, through a "More from Less" approach'.[10]

Needless to say, the next decade will see many innovations in the way agriculture is practiced. From vertical farming to aeroponics, from gene-editing to biopharming; from precision farming to GIS/GPS and AI-enabled tracking and tracing; biologicals to microbiomes for crop protection, and breeding 'climate smart' and 'nutra-rich' crop varieties, frugal solutions will be brought in. Considerable progress has been made by independent innovators in many fields, but a more focused approach, research funding and policy support will be needed in the next decade. Considering that investment in scientific research records the highest returns in agriculture,[11] a substantial rise in funding agricultural R & D will help achieve sustainability through prosperity in agriculture sector, the backbone of the Indian economy.

Firms engaged in the food business will need to think frugal in terms of use of non-conventional crops for nutritious ingredients that are affordable, healthy and can be grown in a sustainable manner. Crops that consume less water and do not damage soil health can be promoted. For example, millet have shown great promise in terms of their nutritional value at much lower costs than wheat flour. They are rich in minerals like calcium, copper, iron, magnesium, phosphorus, potassium, and selenium as well as essential vitamins like folate, pantothenic acid, niacin and riboflavin, and vitamins B6, C, E, and K. Further, while it takes 2500 litres of water to grow 1 kg of rice, only 250–300 litres of water is required to grow 1 kg of millet. Likewise, immunity-boosting super foods that are cheap and readily available, such as flax seeds and minor millets, will also find greater use in a variety of ready-to-eat food meant for BOP communities.

To serve the needs of those with seasonal occupation, families of workers engaged at construction sites and traders living without family support, innovative technology can be developed to create cheap, healthy and ready-to-eat meals that can be purchased from grocery stores and would not require refrigerated storage.

Digital farming is the way sustainable agriculture will be practiced by 2030. Precision farming and smartphone-based application of ICT and big data analytics will be integral to the management of farm activities and development of frugal innovations in agricultural demand forecasting, planning and marketing. Progress made in allied fields such as converting farm waste to wealth, affordable solar panels and storage batteries, innovations in packaging industry and novel use of bio-produce like ethanol from maize and sweet sorghum will all contribute towards future growth of agriculture.

Affordable Education For All

There is no denying the need to strengthen our education system. But the question one must ask is what kind and level of education, and for what purpose. A tribal girl from Odisha, working as a domestic helper in a metro city, spends her hard-earned money to educate her younger siblings. Since there is no high school for girls in her village, she admits her younger sister to a missionary school about 20 km away from her home that has a hostel. On completing senior secondary schooling, she calls her younger sister to the city, hoping that she will get a job, and live a more respectful life than hers. However, the best offer promises her younger sister a daily wage of Rs 160 after 8 hours of back-breaking work in the oppressing work environment of a factory. Realizing that as a domestic helper she gets a monthly salary of Rs 10,000 plus one proper meal, her younger sister accepts. So, apparently education did not improve their lives in any way. There are hundreds of thousands of such young, disillusioned boys and girls, who are eager to train and work hard, but do not gain from education. Little importance is attached to 'learning' in our education system, especially at schools. As a result, classroom education is only seen as a means to receive better employment, which mostly remains unfulfilled, though some indirect benefits are gained, such as appreciation and aptitude to adopt modern technologies, from mobile apps to farming. The NEP, 2020, recognizes the urgent need to bring pedagogical changes and integrate vocational trainings to every stage of schooling. Now that there are primary schools in every village within a kilometre of any household, India needs a new model of learner-centric education and skill development. As observed by A. Banerjee and E. Duflo (2011), if there is real demand for skill, a demand for education will naturally emerge and supply will follow.[12] The education sector needs more delivery and system innovations to gain from technological advancements. By combining modern educational tools and resources (e.g., digital learning, virtual classrooms, interactive holograms, etc.,) from private partners with existing infrastructure owned by the government, better education and learning options can be offered at a fraction of the current cost, even in most disadvantaged regions.

App-based learning solutions can be particularly useful in distant and disadvantaged places. New technologies can revolutionize the experience of learning and utility of basic as well as skill-oriented certification programmes, which will help improve the employability as well as quality of manpower entering the workforce. Strong policy support linking 'Skill India' programmes with virtual learning and industry training will be needed. Online tutorial sites such as Byju's, upGrad, Coursera, etc., have already made their mark. These can reshape the way education is delivered. However, online learning may not suit everyone and high-quality higher education for the masses needs to also be delivered offline. With the participation of NGOs, social and charitable trusts in education, and partnerships with reputed firms (under their CSR programme) for experiential learning, affordable and quality education can be a reality. The new NEP proposes various means through which such changes can be brought in.[13] In addition to the direct benefits of education, indirect benefits, in undertaking a variety of vocations more efficiently, cannot be overlooked.

Affordable Transportation and Safe Mobility

The depleting stock of fossil fuels and alarming levels of pollution in most cities of the world have given impetus to the need for environment-friendly transportation solutions that are affordable (Also see Mission Innovation, chapter 7).

Considerable progress has already been made in manufacturing e-vehicles; however, affordable energy storage is one of the most crucial components, where India still depends on imports of battery packs and cells. Innovations integrating renewable energy with distribution and transmission grids, setting up rural micro-grids with diversified loads and developing storage components are needed to sustain India's transition from fossil fuels to renewable energy use. By using skilled human power and adopting a frugal innovation approach, India can aim to become a hub for battery manufacturing, seizing an opportunity for creating markets and employment. To keep pace with the changing dynamics of energy use, India is now working towards a National Energy Storage Mission (NESM), partnering with stakeholders from all sectors.[14]

The other challenge is in connecting consumers staying in far-flung areas with affordable transport hubs. Micro-transport solutions that leverage environment-friendly technologies are the need of the hour. For example, in the hinterlands of India, Tata Ace, a mini truck, introduced for goods transport, has become a people's carrier in the absence of reliable public transport. Electrified versions of similar vehicles can be deployed in close coordination with local civic authorities to complement the public transport system and also ensure low-cost transportation for BOP consumers. The CSR funds of local corporate firms can be tapped to build the charging infrastructure required to run minivans.

Affordable Microcredit through Easy Access

The importance of access to banking and non-banking microfinancing services for economic security of the people at BOP is now well established. The Government of India's initiative of providing access to banking services to over 30 crore poor people through Jan Dhan accounts is a great step forward, which could be used more effectively by adopting innovative banking services. There are many models available today, such as peer-to-peer lending and online platforms for short-term loans at affordable interest rates. Presently, poor people are paying much higher rates of interest on unsecured loans taken from unauthorized moneylenders or gold loans offered by non-banking finance companies. Some firms like www.lendbox.in have taken the lead in creating online platforms to help people take loans from other people. However, the challenge is to help poor people who may not enjoy access to the Internet or are not well-educated to avail microcredit.

Banking apps and non-traditional banking services, such as digital banks like Simple, SoFi, etc., already having become popular in Europe and now in the US, also stand a chance to become prevalent in India. These app-enabled models, which have features such as quick payments between friends and family, digital cheque deposits via photograph and instant card locking, are expected to become popular with the growing younger generation. Add-ons like automated budgeting tools, investment options, savings goals or tracking everyday expenses can also be introduced.

In all the priorities mentioned above, and many innovative solutions suggested, use of supportive software will play a key role. The Indian software industry has made a mark in the global IT industry and is estimated to export software worth $200 billion (more than Rs 14 lakh crore) annually. The prowess can be leveraged for developing frugal solutions to the problems faced by India. A case in point is the Passport Seva Kendras that are being managed by TCS. Similarly, by roping in the expertise of Indian software firms, the government can build easy-to-access solutions for Indian citizens particularly targeted at BOP communities. The problems that can be solved using software expertise need to be identified, and suitable resources can be allocated to find solutions through software-enabled applications.

The focus should be on finding quick solutions to emergent problems because delay in many cases nullifies the basic purpose of offering a solution, such as a crucial medical situation or expert guidance on fixing a machine that breaks down in the field, particularly to the people at BOP. A good example is the Aarogya Setu app, launched by the Indian government for containment of the Covid-19 virus, by tracking the location of every Covid-19 positive person and providing this information in real time through a mobile app.

Virtual modes of interactions have become, even if temporarily, the new normal in every sphere of life from classrooms to doctors' clinics; from business meetings to banking and a hundred more. However, considering the prevailing situation, a phygital approach (physical-cum-digital) will, perhaps, be more effective for another five–seven years in most cases, rather than digital-only approaches. At the same time, more technological innovations will be expected in the coming years to make virtual experiences more life-like.

Doing Good *Can* Be Good Business Too!

Firms must undertake 'sustainability audits' in a regular manner. The aim ought not to be to just comply with environmental norms (e.g., managing carbon emissions), but proactively focus on reduction/elimination of wastage, replacement of unsustainable materials in the manufacturing process, better social capital, and employee and customer wellness. Waste reduction needs to permeate every aspect of business and not just the materials consumed in the production process. For example, unconsumed food wasted daily in canteens serving factory workers. Most importantly, the goal of sustainable practice is to also help firms achieve higher profits in the long term. For example, switching over to the use of solar power in an office needs to be seen as savings in energy expenses over the next few decades.

Firms can leverage modern trends and offer solutions using frugal innovations. The desire to not own assets is a trend that is catching on with today's young generation. Examples like Oyo Living have shown that there is a demand for minuscule living spaces at very low cost. In the future, uniquely created living spaces can be offered that are compact, digitally connected, automation-enabled, safe, as well as enjoyable to stay. The people living in such small homes can avail of other facilities (e.g., fitness centres, gardens, swimming pools, etc.) through suitable memberships. In urban cities, low-cost living spaces can be extremely beneficial for people engaged in humble jobs at modest salaries.

Innovative business models can help create profits for firms in areas that were previously frowned upon by entrepreneurs. Examples include waste management, sanitation, caring for elders, preventing child trafficking, etc. In most cases, it is an ingenuous use of technology and common sense that can work wonders. Social organizations like the Bill and Melinda Gates Foundation can be roped in as partners to ensure easy access to funds and other resources. Traditionally, businesses shy away from partnering with social organizations. Putting technology to good use will help all stakeholders. Firms can use blockchain technology to not only ensure secure and faster cross-border transactions but also enable people to track the source of transactions easily. This can help consumers obtain additional information about a firm's credentials before engaging in transactions. Furthermore, shady deals and terrorist activities will get caught quickly and the agencies responsible for their neutralization alerted well in time.

In a consumerist society, materialism is the holy grail and movies certainly have perpetuated the belief that it is an I-me-my world, where acquiring wealth and buying possessions are the ultimate achievements. But in today's troubled times, luxury is attaining the garb of sustainability and helping patrons feel as though they are giving back whenever they shop. Brands like Good Earth[15] are promoting sustainable fashion wear. Where's the opportunity for frugal innovation in fashion wear? There are a host of sites that offer fashion wear on rent for a fraction of the cost of the clothing. Recycle, reuse, redesign and recirculate are going to be common practices in the innovative world of frugal fashion in the coming years. Innovative technologies that convert agricultural waste into alternative materials for wearable textile like AltMat are already gaining support from the environment-conscious people.

One of the major concerns of any business is the high cost of setting it up, the high running costs and the high budgets required for brand promotion and distribution. Frugal innovators can leverage a lot of freely available resources for setting up, running the business (outsourcing), and clever use of social media for promotions and distribution. Partnering will help drive down the costs and an asset-light model, like Uber, may become the norm in many industries.

Towards a Better Tomorrow

By leveraging good technology, what seemed impossible yesterday can be made feasible. It only takes imagination to dream up solutions and put them to good use. Simultaneously, the focus needs to be on driving down costs through a variety of ways. The magic of science can help overcome obstacles posed by social evils, regressive cultural beliefs/practices, and enable equitable access to good opportunities.

Frugal innovations can foster good governance. In recent times, several reports in the media have highlighted the shortfalls of agencies involved in delivery of various social initiatives, such as orphanages for girls, mid-day meal schemes in rural schools, pilferage of benefits intended for the poor, etc. It is time for the government to promote simple and efficient innovations to keep a stringent tab on malpractices and put in place technology-enabled early warning signal systems so that such wrongdoings can be nipped in the bud. The use of AI, digital connectivity and real-time monitoring at say anganwadi centres will surely help the Union Ministry of Women and Child Development ensure better delivery schemes. Fortunately, there are many tech-savvy non-profit organizations that are coming forward to create new solutions based on frugal models. For example, the Wadhwani Institute for Artificial Intelligence[16] has created an AI-powered smartphone-based anthropometry tool that will empower health workers to screen low-birth-weight babies without any specialized equipment or physical examinations. The importance of this tool can be understood from the fact that over 20 million (2 crore) newborns worldwide face a life of prolonged and serious ill-health due to their low birthweight (less than 2.5 kg). Babies with low birth weight are more likely to die during their first month of their life and those that manage to survive face lifelong consequences, such as a higher risk of stunted growth, lower IQ, and adult-onset chronic conditions such as obesity and diabetes.

The time has come for resource-deficient and overpopulated nations like India to enable development of solutions that are simple, affordable and accessible. Whether it's the requirement of new/better products for consumption, need to address social challenges or to develop effective processes for governance, innovations alone can show us the way forward. While the market will drive the demand for innovations, the success of the developed solutions will squarely depend on their acceptability by the consumers. And, to ensure equitable distribution of goods and services, frugal innovations alone can be expected to deliver on the promise. Irrespective of the severity of the challenge, the solution for it ought to be sound, robust and fail-safe. To be able to come close to market expectations or to a gold standard implies a high level of technical refinement in frugal innovations, particularly product offerings. The innovation has to qualify on all scientific parameters. Its design and manufacturability should be practical. And the technical specifications and operability should be credible even to the competition to obviate any unfair discrimination or roadblocks in the market place. Such technically sound products are likely to appeal to the early adopters of the innovation. Similarly, frugal innovations for providing services to masses, whether at the lower economic strata or to other disadvantaged sections, should be appropriately designed to integrate with our social values and cultural ethos. Interestingly, there is increasing acceptance of new technologies like AI, machine learning (ML), IoT, the cloud, robotics, 3-D printing, data analytics, nanomaterials, etc., in the development paradigm. And, thanks to the Internet, the new technology tools can be effectively and economically leveraged to deploy various services, whether in agriculture, healthcare or education. While government programmes must support with enabling policies, test beds and adequate development funding, other partners like industry, academia, NGOs, etc., should complement the efforts in the R & D ecosystem, where passionate innovators and willing institutions have embedded the mission of frugal innovations in the philosophy of sustainable growth for all.

Conclusion

Twenty-one years ago, in the foreword of Tom Kelly and Jonathan Littman's international bestseller, *Art of Innovation*, Thomas J. Peters had ascribed the process of successful innovation to the creation of energetic—'hot' in his words— teams, the ability to see the product from a customer's point of view, brainstorming to come up with the best out-of-the-box ideas, followed by rapid prototyping, pilot testing and commercial rollout. These processes need to be interwoven in a precise and imaginative manner to provide an apt solution to a problem which makes this exercise as subtle as an art form.

While planning this book, our idea was to highlight the importance of S&T backstopping frugal innovations. However, as we delved deeper into the subject, we realized that in order to find simple solutions to problems—small and big with minimal use of resources and to make it useful to more people in terms of affordability—the ability to imagine a practical solution for the problem is as vital as giving it a sound scientific base.

Going through examples and stories of various innovators, many of which are discussed in the book, be it Prajapati (creator of Mitticool), Dr Rao (who created Aum) or Dr Sethi and Sharma (who invented the Jaipur Foot), or the highly qualified scientific teams of S&T institutions that developed affordable solutions to the problems of millions. In areas like sustainable agriculture (examples of IARI), tools to fight air pollution, eco-friendly sanitation, affordable medical diagnostics (created at IITs), or waste management, frugal innovations have made a mark as a result of creativity, and the strong desire to solve personal or universal problems which were the guiding force behind all innovation. Similarly, Dr Prakash's origami-based laboratory microscope, Foldscope, and Dr Bhamla's Paperfuge are excellent examples of frugal scientific innovation.

Success of frugal innovation depends on many factors. It starts with a deep sense of empathy or real-life experience of a difficult situation, identifying the crux of the problem and visualizing an imaginary solution. This is often followed by brainstorming on the various issues related to the possible solution particularly with respect to the context of the user, the prevailing social milieu and the usability of science-based tools and technologies with respect to the cost of usage. However, the path from ideation to working out a solution requires thorough technical knowledge of every component that circumscribes the problem. Validation of the innovation, a logical plan to upscale and its largescale diffusion are the final stages of developing a frugal innovation.

Let us conclude with our last story, which will convince you that frugal innovation stands on a sound scientific base nurtured by the sensitivity of an artistic creator.

Senthilkumar Murugesan of Madurai is passionate about innovating through technology to solve real-life problems. The co-founder of JioVio Health Care, Murugesan was shocked to discover that his sister and many women like her who lived on the outskirts of Madurai could not go for regular prenatal check-ups due to distance from their consulting doctors and difficulty in travelling, which resulted in serious health problems for both the mothers and their newborns. After discussing with a consulting doctor, he realized that the crux of the problem lay in monitoring the six vital parameters of expectant mothers and their babies. An engineer by training, he explored applying modern technological solutions to solve this age-old problem. He developed Savemom, an IoT-based maternal healthcare solution that monitors the mother's health using smart wearables (resembling an ornament) that collect six vital physiological signals (blood pressure, heart rate, temperature, respiratory rate, ECG, oxygen saturation and glucose) continuously. These signals are processed continuously for risk assessment and uploaded to the cloud for sharing with the doctors (and others concerned). Necessary medical advice is given to maintain safe health of the mothers. This device, including a battery with a life of six months, costs Rs 1000. The cost also includes fifteen prenatal and antenatal check-ups covering a period of 1000 days! It is not surprising that Murugesan's Savemom bagged the coveted Anjani Mashelkar India Innovation Award in 2020.[1]

The history of humankind is full of such examples, where innovative solutions using imagination and intellect were found for its innumerable and ever-emerging problems. In a world challenged by fast-depleting resources and the multiplier effect of newer and more complex problems, frugal innovations will become more relevant in the days to come.

Acknowledgements

The idea to write this book was seeded during an informal discussion between Malavika and Roshini, the former associate editor at Penguin Random House India, on the importance of scientific institutions in providing affordable technological solutions to the masses. Hence, a lot was discussed about frugal innovations and frugal science, an aspect that is not talked about much. It was her conviction that convinced us to write this book. So, thanks Roshini!

At that point we had no idea how hard it was going to be to converge the views of three science, technology and innovation professionals from diverse fields. We kept changing our narrative, from purely scientific to different levels of everyday science, and finally brought out a book about merging imagination and scientific backstopping to create innovative solutions to everyday problems. In doing so, we became very demanding and even critical of each other at times. We thank each other for staying patient and understanding!

Our special thanks to Dr R.A. Mashelkar, FRS; former DG, CSIR; and past president of Indian National Science

Academy (INSA), who, in spite of his extremely busy schedule, took time for a deep and insightful discussion with Malavika about the relevance of frugal innovations in today's world and the role of science in it. We express our gratitude to him for accepting our request to write the preface of this book.

We would like to express our sincere appreciation for Dr T. Mohapatra, director-general, ICAR; and secretary, Department of Agricultural Research and Education (DARE), Government of India; and president, National Academy of Agricultural Sciences (NAAS), for his support and keen interest in the subject. We also thank Dr R.S. Paroda, president, Trust for Advancement of Agricultural Sciences (TAAS), and past president, NAAS, for providing a lot of useful information on recent scientific and technological innovations in the field of agriculture and Prof. Anil Gupta of Honeybee Foundation at IIM Ahmedabad for sharing his vast experience on grassroot innovations.

Information provided by a number of science professionals from CSIR, ICAR, DBT institutions, IITs and Indian Institute of Science Education and Research, (IISER),Pune, and a number of independent innovators, whose works have been cited in this book, are gratefully acknowledged. Special thanks to the discussants and participants at the event organized in Bengaluru by DIALOGUE: Science, Scientists and Society, Indian Academy of Sciences (IASc), on 'Institutions and Innovations—What Can Research and Educational Institutions Do to Nurture Innovation?' for presenting views from a diverse cross section of academicians, researchers, research managers and inventors.

We would like to express our sincere appreciation for Dr Kheya Bhattacharya, former Indian ambassador to Morocco, who provided valuable insight and information, especially on international and intergovernmental science, technology and innovation policy initiatives, particularly with reference to frugal innovations.

We thank all our colleagues (past and present) and friends in our respective peer groups (difficult to name each of them), who shared their experiences freely and provided valuable inputs, adding value to our efforts.

Our heartfelt thanks to the whole team at Penguin Random House India, especially Milee Ashwarya, Nicholas Rixon and Radhika, who kept our morale high through the tough times during the Covid-19 pandemic and helped shape our efforts into a book that presents a different perspective on frugal innovation.

In the end, we thank our family members, without whose support this book would not have been possible.

Kaushik expresses special thanks to his wife, Reshmi, and daughter, Ritika.

Anil thanks his wife, Rashmi, son, Shivashish, and daughter, Ridhima, for their unswerving support.

Malavika is indebted to Narendra for his unlimited patience and unwavering support, to Partha and Rupa for being her inspiration, and to Deepika for keeping her motivated even at low tide!

In the end, we want to put on record our appreciation for each one who made this book happen.

Malavika Dadlani
Anil Wali
Kaushik Mukerjee

Notes

Preface

1. Defined by Wikipedia as 'an author-level metric that measures both the productivity and citation impact of the publications'.

Introduction

1. Rajnish Tiwari, Luise Fischer and Katharina Kalogerakis, 'Frugal Innovation: An Assessment of Scholarly Discourse, Trends and Potential Societal Implications' in *Lead Market India*, Cornelius Herstatt and Rajnish Tiwari (eds.), pp. 13–35, Springer, Switzerland, 2017.
2. Eric Broudy, *The Book of Looms: A History of the Handloom from Ancient Times to the Present*, Providence: Brown UP, 1979, as quoted by Kristy Beauchesne, Sun Eoh and Kate McClosky in the Museum Directory of Smith College History of Science, https://www.smith.edu/hsc/museum/ancient_inventions/hsc00b.htm (accessed on 19 September 2021).
3. See https://www.history.com/news/sumerians-inventions-mesopotamia.

4. Coimbatore K. Prahalad, *The Fortune at the Bottom of the Pyramid, Revised and Updated 5th Anniversary Edition: Eradicating Poverty Through Profits*, Prentice Hall Publishing, New Jersey, 2009.

5. Navi Radjou and Jaideep Prabhu, *Do Better with Less: Frugal Innovation for Sustainable Growth*, PRHI, Gurgaon, 2019.

6. Abiodun Adegbile and David Sarpong, 'Disruptive Innovation at the Base-of-the-Pyramid: Opportunities, and Challenges for Multinationals in African Emerging Markets', *Critical Perspectives on International Business*, 2018, https://doi.org/10.1108/cpoib-11-2016-0053 (accessed on 1 April 2020).

7. Marco Zeschky, Bastian Widenmayer and Oliver Gassmann, 'Organising for Reverse Innovation in Western MNCs: The Role of Frugal Product Innovation Capabilities', *International Journal of Technology Management*, 64.2–4, pp. 255–275, 2014.

8. Kazuhiro Asakawa, Alvaro Cuervo-Cazurra, and C. Annique Un, 'Frugality-Based Advantage', *Long Range Planning*, 52.4:101879, 2019.

9. Navi Radjou, Jaideep Prabhu and Simone Ahuja, *Jugaad Innovation: Think Frugal, Be Flexible, Generate Breakthrough Growth*, John Wiley & Sons, 2012.

Chapter 1: Frugal Innovation—Emerging Perspectives

1. See 'First Break the Rules: The Charms of Frugal Innovation', *The Economist*, 15 April 2007, https://www.economist.com/node/15879359.

2. Anne-Christin Lehner and Jurgen Gausemeier, 'A Pattern Based Approach to the Development of Frugal Innovations', *Technology Innovation Management Review*, 6.3, pp. 13–21, 2016.

3. Pavan Soni and Rishikesha T. Krishnan, 'Frugal Innovation: Aligning Theory, Practice, and Public Policy', *Journal of Indian Business Research*, 6.1, pp. 29–47, 2014.

4. See https://www.electricaltechnology.org/2015/04/types-of-ics-classification-of-integrated.

5. 'Frugal Innovation Evolves in the Next Phase of China's Rise as Tech Economy', *Forbes*, 19 November 2013.

6. Navi Radjou and Jaideep Prabhu, *Frugal Innovation: How to do More with Less,* Economist Books, New York, 2015.

7. Navi Radjou and Jaideep Prabhu, *Do Better with Less,* PRHI, Gurgaon, 2019.

8. R.A. Mashelkar, 'Building a Great Indian Agricultural Innovation System', convocation address at Mahatma Phule Krishi Vidyapeeth, Rahuri, Maharashtra, 28 February 2015.

9. Marco B. Zeschky, Stephan Winterhalter and Oliver Gassmann, 'From cost to frugal and reverse innovation: Mapping the field and implications for global competitiveness', *Research-Technology Management,* 57.4, pp. 20–27, 2014.

10. P. Basu, 'Science on a Shoestring: Microscopes made from Bamboo bring Biology into Focus, *Nature Medicine,* 13.10, pp. 1128–1129, 2007.

11. Balkrishna, C. Rao, 'Advances in Science and Technology Through Frugality', *IEEE Engineering Management Review,* 45.1, pp. 32–38, 2017.

12. Coimbatore K. Prahalad and Raghunath A. Mashelkar, 'Innovation's Holy Grail', *Harvard Business Review,* 88.7–8: 132–141, 2010.

13. R.A. Mashelkar, 'Innovation's Holy Grail—more from less for more', www.idrbt.ac.in/assets/pdf/FD_2013/RAM_IDRBT.pdf.

14. G. Natchiar, R.D. Thulasiraj and R. Meenakshi Sundaram, 'Cataract Surgery at Aravind Eye Hospitals: 1988–2008', *Community Eye Health,* 21.67, p. 40, 2008.

15. T. Khanna and T. Bijlani, 'Narayana Hrudayalaya Heart Hospital: Cardiac Care for the poor (B)', *Harvard Business School Strategy Unit Case,* p. 712, 2011.

16. New Innovation approaches to support implementation of Sustainable Development Goals, UNCTAD/DTL/STIC/2017/4, Copyright UN, 2017.

17. See Margaret Molloy, *Insider,* 25 December 2013, https://thenextweb.com/insider/2013/12/25/simplicity-sells-10-surprising-innovations-make-life-simpler/.

18. R. Meierhofer and M. Wegelin, 'Solar water disinfection—A guide for the application of SODIS', *Swiss Federal Institute of Environmental Science and Technology (EAWAG) Department of Water and Sanitation in Developing Countries (SANDEC)*, 2002, https://en.wikipedia.org/wiki/Solar_water_disinfection.

19. See https://akvppedia.org/wiki/uv-treatment-solar-disinfection-sodis; https://www.practo.com/healthfeed/harmful-effects-of-drinking-water-in-plastic-bottles-4343/post?utm_source=Consult&utm_medium=related_articles.

20. Less than two million people suffer from movement disability as per Census, 2011/updated 2016, https://enabled.in/wp/census-of-india-2011-disabled-population/.

21. A.K. Jain and M. Kejariwal, 'Biodegradability and Waste Management of Jaipur Foot', *PharmaTutor*, 4.2, pp. 41–44, 2016, https://www.pharmatutor.org/magazines/articles/february-2016/biodegradability-waste-management-jaipur-foot (accessed on 18 August 2019); Also see, https://arthaimpact.com/latest_news/jaipur-foot-one-of-the-most-technologically-advanced-social-enterprises-in-the-world/ and https://www.jaipurfoot.org/how-we-do/research-and-development.html.

22. Ibid.

23. See https://www.gsb.stanford.edu/faculty-research/publications/jaipurknee-project-ii-scaling-business.

24. See http://www.forbes.com/sites/rahimkanani/2011/08/08/Jaipur-foot-one-of-the-most-tecgnologically-advanced-social-enerprises-in-the-world/.

25. See https://www.infosys.com/infosys-foundation/aarohan-social-innovation-awards.

26. Eureka, interview with T.V. Venkateswaran on Rajya Sabha TV, 7 October 2017.

27. Dr Dhananjaya Dendukuri at 'Innovation and Institutions', a panel discussion organized by the Indian Academy of Sciences at Indian Institute of Sciences (IISc), Bengaluru, on 10 August 2019, www.achiralabs.com.

28. See https://www.hhmi.org/stories/qa-manu-prakash-philosophy-frugal-science.

29. Manu Prakash, 'Six Lessons I Wish I Had Learned Sooner', IndiaBioSience.com.
30. J. S. Cybulski, J. Clements and M. Prakash, 'Foldscope: Origami-Based Paper Microscope', *PLoS ONE*, 9.6, p. e98781, 2014, https://doi.org/10.1371/journal.pone.0098781.
31. 'Frugal Science: What, Why and How', https://www.janelia.org/past-dialogues-discovery-lectures/manu-prakash.
32. See web.stanford.edu/group/prakash-lab/AnnualReport2015.pdf.
33. Saad, M. Bhamla, Brandon Benson, Chew Chai, Georgios Katsikis, Aanchal Johri and Manu Prakash, 'Hand-Powered Ultralow-Cost Paper Centrifuge', *Nature Biomedical Engineering*, 1.0009, 2017, https://www.nature.com/articles/s41551-016-0009.
34. Discussion with Malavika Dadlani on 7 April 2020.
35. See www.biospectrumindia.com/news/57/11669/researchers-in-canada-design-wearable-ultrasound-scanner.html
36. R.N. Basu, 'An Appraisal of Research on Wet and Dry Physiological Seed Treatments and Their Applicability with Special Reference to Tropical and Sub-Tropical Countries', *Seed Science and Technology*, 22.1, pp. 107–126, 1994.
37. R.N. Basu and M. Dasgupta, 'Control of Seed Deterioration by Free Radical Controlling Agents', *Indian Journal of Experimental Biology*, 16, pp. 1070–1073, 1978.
38. Discussion with the late C. Dharmalingam, former professor and head, seed technology, TNAU, Coimbatore.
39. P.V. Manoranjan Rao, B.N. Suresh and V.P. Balagangadharan, *From Fishing Hamlet to Red Planet: India's Space Journey*, HarperCollins, 2015.
40. See 'The ISRO isn't enough. India needs its own Elon Musk or Jeff Bezos', Rajeswari Pillai Rajagopalan, Quartz India, 19 July 2019, https://qz.com/india/1668876/indias-quest-to-take-the-isro-to-the-top.
41. 'South American countries look towards India for low-cost satellite launches', *Financial Express*, 12 March 2019, https://www.financialexpress.com/lifestyle/science/isros-big-achievement-south-american-countries-look-towards-india-for-low-cost-satellite-launches/1513776/.

42. E. Rosca, J. Reedy and J.C. Bendul, 'Does Frugal Innovation Enable Sustainable Development? A Systematic Literature Review', *The European Journal of Development Research*, 30.1, pp. 136–157, 2017, https://link.springer.com/article/10.1057/s41287-017-0106-3.
43. Mario Pansera, 'Frugal or Fair? The Unfulfilled Promises of Frugal Innovation', *Technology Innovation Management Review*, 8.4, pp. 6–13, 2018.
44. See https://www.jstor.org/stable/23240265.

Chapter 2: Science in Frugal Innovation

1. See https://innovationleadershipforum.org/our-wisdom/why-do-we-innovate.
2. See https://emps.exeter.ac.uk/physics-astronomy/staff/ns329.
3. Pisoni, Alessia, Laura Michelini, and Gloria Martignoni, 'Frugal approach to innovation: State of the art and future perspectives', *Journal of Cleaner Production*, 171, pp. 107–126, 2018.
4. Coimbatore K. Prahalad, 'Bottom of the Pyramid as a Source of Breakthrough Innovations', *Journal of Product Innovation Management*, 29.1, pp. 6–12, 2012.
5. P. Soni and T. Krishnan, R., 'Frugal Innovation: Aligning Theory, Practice, and Public Policy', *Journal of Indian Business Research*, 6.1, pp. 29–47, 2014.
6. 'How 3 Indian Doctors Pioneered the Use of ORS to Treat Diarrhoea & Saved Millions', Better India, 17 October 2019, https://www.thebetterindia.com/200457/forgotten-bengal-doctor-pioneered-ors-treatment-diarrhoea-cholera-saved-millions-india/.
7. Joshua N. Ruxin, 'Magic Bullet: The History of Oral Rehydration Therapy', *Medical History*, 38.4, pp. 363–397, 1994.
8. See https://www.thebetterindia.com/190240/shrawan-kumar-rajasthan-sikar-farmer-innovations.
9. See http://nif.org.in/upload/news/9th-National-Biennial-Grassroots-Innovation-Awards-2017.pdf.

10. S. Najafi-Tavani, Z. Najafi-Tavani, P. Naude, P. Oghazi and E. Zeynaloo, 'How Collaborative Innovation Networks Affect New Product Performance: Product Innovation Capability, Process Innovation Capability, and Absorptive Capacity', *Industrial Marketing Management*, 73, pp. 193–205, 2018.

11. Jorg Thoma, 'DUI mode learning and barriers to innovation—A case from Germany', *Research Policy*, 46.7, pp. 1327-1339, 2017.

12. Harvey Brooks, 'The Relationship Between Science and Technology', *Research Policy*, 23.5, pp. 477–486, 1994.

13. Food and Agriculture: Technology Vision 2020, TIFAC, 1998, see http://tifac.org.in.

14. Discussion with R.K. Pal, a member of Dr Roy's team, on 21 November 2019.

15. See Better India, 10 September 2019, www.thebetterindia.com/194656/gurtej-sandhu-indian-origin-inventor-more-patents-than-edison-inspiring/.

16. Balakrishna C. Rao, 'Science Is Indispensable to Frugal Innovations'. *Technology Innovation Management Review*, 8.4, pp. 49–56, 2018.

17. Anil Gupta, 'From Sink to Source—the Honeybee Network Documents Indigenous Knowledge and Innovation in India', *Innovations*, Summer/2006, pp. 49–66.

18. Application number 201811034565, 2018.

19. GI number 145 of the GI Registry, Government of India, vide certificate number 238, dated 15 February 2016.

20. E.A. Siddiq, L.R. Vemireddy and J. Nagaraju, 'Basmati Rices: Genetics, Breeding and Trade', *Agricultural Research*, 1, pp. 25–36, 2012.

21. Vijaipal Singh, Ashok K. Singh, Trilochan Mohapatra, Gopala Krishnan S. and Ranjith K. Ellur, 'Pusa Basmati 1121—a Rice Variety with Exceptional Kernel Elongation and Volume Expansion After Cooking', *The Rice Journal*, 11, p. 19, 2018, https://thericejournal.springeropen.com/articles/10.1186/s12284-018-0213-6.

22. ICAR-IARI Technologies for Farmers' Prosperity, 2018, published by ICAR-IARI, New Delhi.

23. Discussion with Dr A.K. Singh, director, ICAR-IARI, on 21 March 2020.

24. See http://agriexchange.apeda.gov.in/indexp/Product_description_32headChart.aspx?gcode=0601.

25. See *Hindustan Times*, 7 March, 2017; https://thelogicalindian.com/exclusive/akash-manoj/?infinitescroll=1.15 November 2017.

26. See 'First break all the rules: The Charms of Frugal Innovation', *The Economist*, 15 April 2010, https://www.economist.com/node/15879359.

27. P. Basu, 'Science on a Shoestring: Microscopes Made from Bamboo Bring Biology into Focus', *Nature Medicine*, 13.10, pp. 1128–1129, 2007.

28. B.C. Rao, 'Advances in Science and Technology through Frugality', *IEEE Engineering Management Review*, 451, pp. 32–38, 2017, http://doi.org/10.1109/EMR.2017.2667219.

29. See www.downtoearth.org.in/coverage/bullock-cart-in-its-new-avatar-12232.

30. See Larry Myler, 'Innovation Is Problem Solving . . . And A Whole Lot More', *Forbes*, 13 June 2014, https://www.forbes.com.sites/larrymyler/2014/06/13innovation.is.problem.solving.and.a.whole.lot.more.

31. Anirudh Krishna, *The Broken Ladder: The Paradox and The Potential of India's One Billion*, PRHI, Gurgaon, 2018.

32. See www.sristi.org/hbnew/index.php.

33. See sristibionest.wordpress.com/2019/02/03/commercialization.

34. See nif.org.in/awards.

35. Christoph Zott and Raphael Amit,'Business Model Innovation: Creating Value in Times of Change', *Working Paper*, IESE Business School, University of Navarra, 2010.

36. 'Bharat Biotech's new variant of vaccine gets WHO pre-qualification', *The Hindu*, 26 March 2016, https://www.thehindu.com/news/cities/Hyderabad/bharat-biotechs-new-variant-of-vaccine-gets-who-pre-qualification/article35683333.ece

37. Balkrishna, C. Rao, 'Science is Indispensable to Frugal Innovations', *Technology Innovation Management Review*, 8.4, pp. 49–56, 2018.
38. See www.villgro.org.
39. See http://assistech.iitd.ernet.in/smartcane.php.
40. See https://innovation.mit.edu/resource/d-lab/.
41. See https://www.cfia.nl/kenya-hub.
42. See https://www.scu.edu/engineering/labs--research/labs/frugal-innovation-hub/.
43. History of polythene, https://www.globalplasticsheeting.com/our-blog-resource-library/bid/23095/The-History-of-Polyethylene.

Chapter 3: A View of Successes and Failures

1. Coimbatore, K. Prahalad, *The Fortune at the Bottom of the Pyramid*, Wharton School Publishing, 2006.
2. Raghunath Mashelkar and Ravi Pandit, *Leap Frogging to Pole-Vaulting*, PRHI, Gurgaon, 2019.
3. See *Business Standard*, 20 January 2013, www.business-standard.com/article/companies/ge-healthcare-launches-low-cost-portable-ecg09112400032.
4. See https://sanketlife.in/pages/about-us.
5. See press release, The Nobel Prize in Chemistry 2019, www.nobelprize.org/prizes/chemistry/2019/press-release/.
6. *See MIT Technology Review* 2019, https://www.technologyreview.com/2019/02/27/66011/the-10-worst-technologies-of-the-21st-century/.
7. See www.inc.com, 23 March 2019, https://www.inc.com/sean-wise/mit-announced-top-10-worst-tech-innovations-of-21st-century-heres-what-they-all-have-in-common.html
8. Sean Wise, https://www.inc.com/sean-wise/mit-announced-top-10-worst-tech-innovations-of-21st-century-heres-what-they-all-have-in-common.html.
9. Paul Sloane, 'A Lesson in Innovation—Why Did the Segway Fail', *Innovation Management*, 5, 2012; and Raghunath Mashelkar and

Ravi Pandit, *Leap Frogging To Pole-Vaulting*, PRHI, Gurgaon, p. 346, 2019.

10. Harvey Brooks, 'The Relationship Between Science and Technology', *Research Policy*, 23.5, pp. 477–486, 1994.

11. Ashok Gulati, Marco Ferroni and Yuan Zhou, 'Supporting Indian Farms the Smart Way', *Academic Foundation*, p. 456, 2018.

12. Personal interview with T. Mohapatra, director-general, ICAR, 2019.

13. Importance of R & D to innovation, www.incrementalinnovation. com/innovation-management-development/rd-to-innovation.

14. Cry proteins are crystalline proteins which are secreted by the bacteria *Bacillus thuringiensis* (commonly referred as Bt) and are known to be toxic to certain groups of insects. The toxin is coded by a group of homologous cry genes and can be incorporated in many crop plants such as Bt cotton, Bt corn, etc., to provide insect pest resistance.

15. K. Kranthi, 'Testing Seed Quality of Cotton Seed', *Cotton Statistics and News*, 23, pp. 1–3, 2013.

16. See https://www.jknanosolutions.com.

17. See https://medicaldialogues.in/medicine/news/iit-researchers-develop-cost-effective-covid-19-test-64135.

18. See https://timesofindia.indiatimes.com/india/pune-based-firm-develops-testing-kit-for-faster-confirmation-of-covid-19-cases/articleshow/74790407.cms.

19. Ramachandran N., 'India's frugal innovation', https://www.livemint.com/Opinion/rqLQXHErXePWEZhhIzTncI/Indias-frugal-innovation.html.

20. See https://www.coe.gatech.edu/news/2019/06/frugal-science-age-innovation.

21. See http://www.advantagenature.com/press-media/index.html.

22. Ashok N. Bhaskarwar and Edward L. Cussler, 'Pollution-Preventing Lithographic Inks', *Chemical Engineering Science*, 52.19, pp. 3227–3231,1997; Sankar Nair and Ashok N. Bhaskarwar, 'Washing Kinetics of Pollution-Preventing Lithographic Inks', *Chemical Engineering Science*, 55.10, pp. 1921–1923, 2000; C.S. Maji and Ashok N. Bhaskarwar,

'Pollution-Preventing Writing, Stamping, Screen-Printing, and Lithographic Inks, 2006; S. Bhimania, K.G. Singh, S. Mourya and A.N. Bhaskarwar, 'Water Reducible Alkyd Resins', WIPO Patent Application WO/2010/02353, Application number: IB2009/006627, publication date: 4 March 2010, filing date: 24 August 2009, 2010.

23. See https://economictimes.indiatimes.com/power-of-ideas-ennatura-makes-biodegradable-ink-using-vegetable-oils/articleshow/6088472.cms?from=mdr.

24. Personal interview to Malavika Dadlani in 2019.

25. See https://anilg.sristi.org/transforming-indian-villages-through-innovations-knowledge-network-and-entrepreneurship/.

26. Anil K. Gupta, 'From Sink to Source: The Honey Bee Network Documents Indigenous Knowledge and Innovations in India', *Innovations: Technology, Governance, Globalization*, 1.3, pp. 49–66, 2006.

Chapter 4: Frugal Innovation for Sustainable Solutions

1. See https://doordarshan.gov.in/ddkisan/dd-mahila-kisan-awards/.

2. See *Deccan Herald*, 28 July 2013, https://www.deccanherald.com/content/347627/ideas-abyss-need-ground-grow.html; The Better India, 19 February 2020, www.thebetterindia.com.

3. FAO AquaStat, 2011.

4. See https://icar.org.in/content/jalopchar-eco-friendly-wastewater-treatment-technology.

5. Liqa Raschid-Sally and Priyantha Jayakody, 'Drivers and Characteristics of Wastewater Agriculture in Developing Countries: Results from a Global Assessment', 2009, https://www.researchgate.net/publication/242230848_Drivers_and_Characteristics_of_Wastewater_Agriculture_in_Developing_Countries_Results_from_a_Global_Assessment.

6. ICAR-IARI Technologies for Farmers' Prosperity, 2018.

7. See www.wateraidindia.in/sites/g/files/jkxoof336/files/state-of-urban-water-supply.pdf.

8. See https://www.iitm.ac.in/content/iitm-scholars-win-gyti-awards-2018.

9. V.M. Chariar and Er S. Ramesh Sakthivel, *Ecological Sanitation Practitioner's Handbook*, Ecological Sanitation Group of the Centre for Rural Development and Technology, Indian Institute of Technology Delhi and UNICEF, India, 2011.

10. Ramesh, S. Sakthivel, Md Azizurrahaman, V. Ganesh Prabhu and V. M. Chariar, 'Performance Evaluation of a Low-Cost Odour Trap Installed in Waterless Urinals', *Journal of Water, Sanitation and Hygiene for Development*, 6.2, pp. 252–258, 2016.

11. See https://www.waterless.com/blog/why-waterless-urinals-may-be-more-sanitary-than-flush-urinals.

12. See http://www.ecosanres.org/pdf_files/Nanning_PDFs/Eng/Calvert%2014_E02.pdf.

13. See @venture_center on Twitter.

14. BIRAC—Lab to Market, 2019, https://birac.nic.in/webcontent/1553774543_BIRAC_Lab_to_Market_Manual_booklet_28_03_2019.pdf

15. Conformité Européene, which literally means European Conformity.

16. A machine and process patent for manufacturing breast prosthesis was granted; Indian Patent Application 2282/DEL/2015.

17. BIRAC—Lab to Market; http://www.dailygood.org/story/2174/world-class-voice-prosthesis-for-1-usha-prasad/

18. Read more at: https://yourstory.com/2015/12/dr-vishal-rao.

19. Detecting just one cell out of 1,00,0000-10,00,0000 leukocytes of the peripheral blood of cancer patients is akin to searching for a needle in a haystack.

20. See https://www.oncodiscover.com/single-news.aspx?not=8; Also see, https://fit.thequint.com/cancer/pune-man-develops-cheap-cancer-test#read-more.

21. See *The Hindu*, 6 August 2019, www.thehindu.com/society/one-cube-revolution/article28835003.ece.

22. See the *BusinessLine*, 'These urban forests keep India breathing freely', 3 December 2018.

23. Agriculture Grand Challenge—Seeding AgriTech Innovation; https://www.startupindia.gov.in/content/sih/en/bloglist/blogs/Innovations_Agritech_AGC.html.

24. See   https://pib.gov.in/PressReleseDetailm.aspx?PRID=1700749.

25. Nirmalya Kumar and Phanish Puranam, 'Have You Restructured for Global Success?', *Harvard Business Review*, 89.10, p. 123, 2011.

Chapter 5: Creating Value through Frugal Innovations

1. Hareem Arshad, Marija Radić and Dubravko Radić, 'Patterns of Frugal Innovation in Healthcare', *Technology Innovation Management Review*, 8.4, pp. 28–26, 2018.

2. See https://www.fortuneindia.com/enterprise/social-commerce-the-meesho-way/105563.

3. Michele-Lee Moore, Darcy Riddell and Dana Vocisano, 'Scaling Out, Scaling Up, Scaling Deep: Strategies of Non-Profits in Advancing Systemic Social Innovation', *Journal of Corporate Citizenship*, 58, pp. 67–84, 2015.

4. Kevin André and Anne-Claire Pache, 'From Caring Entrepreneur to Caring Enterprise: Addressing the Ethical Challenges of Scaling Up Social Enterprises', *Journal of Business Ethics*, 133.4, pp. 659–675,2016.

5. See https://www.jagran.com/uttar-pradesh/aligarh-city-post-office-will-also-bring-bank-services-with-your-door-18380737.html.

6. Geert Hofstede, 'Motivation, Leadership, and Organization: Do American Theories Apply Abroad?', *Organizational Dynamics*, 9.1, pp. 42–63, 1980.

7. Jennifer S. Labrecque, Wendy Wood, David T. Neal and Nick Harrington, 'Habit Slips: When Consumers Unintentionally Resist New Products', *Journal of the Academy of Marketing Science*, 45.1, pp. 119–133,2017.

8. See kamalkisan.com.

9. See www.ideasforindia.in/topics/environment/happy-seeder-a-solution-to-agricultural-fires-in-north-india.html.

10. Coimbatore K. Prahalad, 'Bottom of the Pyramid as a Source of Breakthrough Innovations', *Journal of Product Innovation Management*, 29.1, pp. 6–12, 2012.

11. Jaques Angot and Loïc Plé, 'Serving Poor People in Rich Countries: The Bottom-of-The-Pyramid Business Model Solution', *Journal of Business Strategy*, 36.2, pp. 3–15, 2015.

12. Fabian Eggers, Thomas Niemand, Matthias Filser, Sascha Kraus and Jennifer Berchtold, 'To Network or Not To Network–Is That Really The Question? The Impact of Networking Intensity and Strategic Orientations on Innovation Success', *Technological Forecasting and Social Change*, 155: 119448, 2020.

13. See https://map.sahapedia.org/article/Samagra-Sanitation/2444.

14. Kim, W. Chan and Renée A. Mauborgne, *Blue Ocean Strategy, Expanded Edition: How to Create Uncontested Market Space and Make the Competition Irrelevant*, Harvard Business Review Press, Massachusetts, 2015.

15. Nathan Furr, Kate O'Keeffe and Jeffrey H. Dyer, *Managing Multiparty Innovation*, Harvard Business Review, 94.11, pp. 76–83, 2016.

16. Marc Pfitzer, Valerie Bockstette and Mike Stamp, 'Innovating for Shared Value', *Harvard Business Review*, 91.9, pp. 100–107, 2013.

17. Michael E. Porter and Mark R. Kramer, 'The Big Idea: Creating Shared Value, Rethinking Capitalism', *Harvard Business Review*, 89.1/2, pp. 62–77, 2011.

18. Balkrishna C. Rao, 'Science is Indispensable to Frugal Innovations', *Technology Innovation Management Review*, 8.4, pp. 49–56, 2018.

19. Arpita Agnihotri, 'Revisiting the Debate over the Bottom of the Pyramid Market', *Journal of Macromarketing*, 32.4, pp. 417–423, 2012.

20. Rosa Caiazza, 'A Cross-National Analysis of Policies Affecting Innovation Diffusion', *The Journal of Technology Transfer*, 41.6, pp. 1406–1419, 2016.

21. Peter Knorringa, Iva Peša, André Leliveld and Cees van Beers, 'Frugal Innovation and Development: Aides or Adversaries?', *The European Journal of Development Research*, 28.2, pp. 143–153, 2016.

22. See https://www.fabindia.com/about-us.

23. Wairokpam Premi Devi and Hemant Kumar, 'Frugal Innovations and Actor–Network Theory: A Case of Bamboo Shoots Processing in Manipur, India', *The European Journal of Development Research*, 30.1, pp. 66–83, 2018.
24. India's CSR Act, 2014.

Chapter 6: Market-driven Solutions

1. See https://www.pkotler.org/quotes-from-pk.
2. Balkrishna C. Rao, 'Science is Indispensable to Frugal Innovations', *Technology Innovation Management Review*, 8.4, pp. 49–56, 2018.
3. Koteshwar Chirumalla, 'Organizing Lessons Learned Practice for Product–Service Innovation', *Journal of Business Research*, 69.11, pp. 4986–4991, 2016.
4. Timothy Keiningham, Lerzan Aksoy, Helen L. Bruce, Fabienne Cadet, Natasha Clennell, Ian R. Hodgkinson and Treasa Kearney, 'Customer experience driven business model innovation', *Journal of Business Research*, 116 (2020): 431-440.
5. David C. Edelman and Marc Singer, 'Competing on Customer Journeys', *Harvard Business Review*, 93.11, pp. 88–100, 2015.
6. See www.medtechglobal.com.
7. Rajibul Hasan, Ben Lowe, and Dan Petrovici, 'Consumer adoption of pro-poor service innovations in subsistence marketplaces', *Journal of Business Research*, 121, 461–475, 2020.
8. www.solarear.com.br
9. Michael Barrett, Elizabeth Davidson, Jaideep Prabhu and Stephen L. Vargo, 'Service Innovation in the Digital Age: Key Contributions and Future Directions', *MIS Quarterly*, 39.1, pp. 135–154, 2015.
10. Carlos Bianchi, M. Bianco, M. Ardanche and M. Schenck, 'Healthcare Frugal Innovation: A Solving Problem Rationale Under Scarcity Conditions', *Technology in Society*, 51, pp. 74–80, 2017.
11. Yoichiro Igarashi and Makoto Okada, 'Social Innovation Through a Dementia Project using Innovation Architecture', *Technological Forecasting and Social Change*, 97, pp. 193–204, 2015.

12. V. Govindarajan, 'A Reverse Innovation Playbook', *Harvard Business Review*, 90.4, pp. 120–125, 2012.
13. Darrell K. Rigby, Jeff Sutherland and Hirotaka Takeuchi, 'Embracing Agile', *Harvard Business Review*, 94.5, pp. 40–50, 2016.
14. Tiziana Casciaro, Amy C. Edmondson and Sujin Jang, 'Cross-Silo Leadership How to Create More Value by Connecting Experts from Inside and Outside the Organization', *Harvard Business Review*, 97.3, pp. 130–139, 2019.
15. Eugenia Rosca, Jack Reedy and Julia C. Bendul, 'Does Frugal Innovation Enable Sustainable Development? A Systematic Literature Review', *The European Journal of Development Research*, 30.1, pp. 136–157, 2018.
16. Jeffrey L. Cummings and Bing-Sheng Teng, 'Transferring R&D Knowledge: The Key Factors Affecting Knowledge Transfer Success', *Journal of Engineering and Technology Management*, 20.1-2, pp. 39–68, 2003.
17. Ronny Reinhardt, Sebastian Gurtner and Abbie Griffin, 'Towards an Adaptive Framework of Low-End Innovation Capability—A Systematic Review and Multiple Case Study Analysis', *Long Range Planning*, 51.5, pp. 770–796, 2018.
18. Navi Radjou and Jaideep Prabhu, *Do Better with Less: Frugal Innovation for Sustainable Growth*, PRHI, Gurgaon, 2019.
19. Mirva Hyypiä and Rakhshanda Khan, 'Overcoming Barriers to Frugal Innovation: Emerging Opportunities for Finnish SMEs in Brazilian Markets', *Technology Innovation Management Review*, 8.4, pp. 38–48, 2018.
20. Sonal H. Singh, Bhaskar Bhowmick, Dale Eesley and Birud Sindhav, 'Grassroots Innovation and Entrepreneurial Success: Is Entrepreneurial Orientation a Missing Link?', *Technological Forecasting and Social Change*, Article no. 119582, 2019.
21. Michael A. Hitt, R. Duane Ireland, David G. Sirmon and Cheryl A. Trahms, 'Strategic Entrepreneurship: Creating Value for Individuals, Organizations, and Society', *Academy of Management Perspectives*, 25.2, pp. 57–75, 2011.
22. Jeanne Liedtka, 'Why Design Thinking Works', *Harvard Business Review*, 96.5, pp. 72–79, 2018.

23. David B. Audretsch and Erik E. Lehmann, 'Does the Knowledge Spillover Theory of Entrepreneurship Hold for Regions?', *Research Policy*, 34.8, pp. 1191–1202, 2005.

24. Charles O'Reilly and Andrew J.M. Binns, 'The Three Stages of Disruptive Innovation: Idea Generation, Incubation and Scaling', *California Management Review*, 61.3, pp. 49–71, 2019.

25. Hanna Nari Kahle, A. Dubiel, H. Ernst and J. Prabhu, 'The Democratizing Effects of Frugal Innovation: Implications for Inclusive Growth and State-Building', *Journal of Indian Business Research*, 5.4, pp. 220–234, 2013.

26. https://www.cgdev.org/publication/blueprint-market-driven-value-based-advance-commitment-tuberculosis.

27. Azhar Mohd, et al., 2020, bioRxiv, https://doi.org/10.1101/2020.04.07.028167.

28. See www.thehindubusinessline.com/news/science/covid-indian-researchers-have-redesigned-feluda-kit-says-report/article34072977.ece.

Chapter 7: National Policies and Role of International Organizations

1. PM Dr Manmohan Singh's address at Indian Science Congress, 2010.

2. Science, Technology and Innovation Policy, 2013; DST, Ministry of S&T, Government of India.

3. Discussion with Prof. Ramesh Chand, member, NITI Aayog, on 30 December 2019.

4. Dayal, 2011; *Business Today*, New Delhi, Updated: 6 June 2014.

5. See https://heavyindustries.gov.in/writereaddata/Content/NEMMP2020.pdf.

6. Mark Schulz, 'Innovation: History's Great Free Lunch', *WIPO Magazine*, June 2017.

7. Danielle Archibugi, Andrea Filippetti and Marlon Frenz, 'The Impact of the Economic Crisis on Innovation: Evidence from Europe', *Technological Forecasting and Social Change*, 80.7, pp. 1247–1260, 2013.

8. Rajnish Tiwari, Luise Fischer and Katharina Kalogerakis, 'Frugal Innovation in Scholarly and Social Discourse: an Assessment of Trends and Potential Societal Implications', in project 'Potentiale, Herausforderungen und gesellschaftliche Relevanz frugaler Iinnovationen in Deutschland im Kontext des globalen Innovationswettebewerbs', supported by BMBF, 2016.

9. Coimbatore K. Prahalad and Stuart L. Hart, *The Fortune at the Bottom of the Pyramid*, Strategy + Business, Booz Allen Hamilton Inc, 2002.

10. L. Klerkx and J. Guimon, 'Attracting Foreign R&D Through International Centres of Excellence: Early Experiences from Chile', *Science and Public Policy*, 44.6, pp. 763–774, 2017.

11. Extract from minutes of meeting of the Council for Trade-related aspects of IPR, WTO, item 12, 'Intellectual Property and Innovation: Cost Effective Innovation'; extracted from Document IP/C/M/73/Add.1

12. UNGA Resolution 72/228: *Science, technology and innovation for . . . https://unctad.org › Publications Library › A_res_72_228_en, 2017.*

13. New Innovation approaches to support implementation of Sustainable Development Goals. UNCTAD/ DTL/ STIC/2017/4, Copyright UN, 2017.

14. UNFCCC Article 21 March 2018, UN Climate Change news, https://unfccc.int/news/finding-ways-to-boost-climate-tech-innovation.

15. UNESCO Science Report: towards 2030 (2015), http://uis.unesco.org/sites/default/files/documents/unesco-science-report-towards-2030-part1.pdf.

16. Vinay Pandey and Ruchika Chitravanshi, 'India is a Leader in Frugal and Demand-Driven Innovation: Francis Gurry', the *Economic Times*, 6 March 2017.

17. Rebecca Fanin, 'Frugal Innovation in the Next Phase of China's Rise as Tech Economy', *Forbes*, 19 November 2013.

18. Arasaratnam Ajanthy and Gary Humphreys, 'Emerging Economies Drive Frugal Innovation', *Bulletin of the WHO*, 91, pp. 1–80, 2013.

19. WHO Bulletin, Volume 91, No. 1, 2013, https://www.who.int/bulletin/volumes/91/1/13-020113.pdf.

20. See businesstoday.in, 'GE launches new low cost infant warmers', *Business Today*, 13 December 2013 (accessed on 6 December 2021).

21. See www.oecd.org/sti/inn/knowledge-and-innovation-for-inclusive-development (accessed 20 November 2019).

22. 2017 UNCTAD study, https://unctad.org/system/files/official-document/dtlstict2017d4_en.pdf.

23. Barefoot grandmothers electrify rural communities, *CNN* interview with Sanjit Roy, 27 January 2011.

24. See www.mds.gov.br/programas/seguranca-alimentar-e-nutricional-san/cisterns.

25. V. Pandey and R. Chitravanshi, 'India is a leader in frugal and demand-driven innovation: Francis Gurry', The *Economic Times*, 6 March 2017.

26. K. Jayaraman, US patent office withdraws patent on Indian herb, *Nature*, 389, 6 (1997); https://doi.org/10.1038/37838.

27. Christian Le Bas, 'Frugal Innovation, Sustainable Innovation, Reverse Innovation: Why Do They Look Alike? Why Are They Different?', *Journal of Innovation Economics Management*, 21.3, pp. 9–26, 2016.

28. Rhythma Kaul, 'India moves up on innovation index, *Hindustan Times*, 25 July 2019, www.hindustantimes.com/india-news/india-moves-up-on-innovation-index/story-jvoYrGRFz9C90Jx6tq4stI.html ; Also see www.wipo.int/edcos/pubdocs/en/wipo_pub_gii_2021.pdf.

29. 'Work for A Brighter Future, Global Commission on the future of work', International Labour Organization, published in January 2019.

30. M. Leach, A. Sumner and L. Waldman, 'Discourses and Disquiet: Multiple Knowledge in Science, Society and Development', *Journal of International Development*, 20.6, pp. 727–738, 2008.

31. See http://mission-innovation.net.

Chapter 8: Science and Sustenance: The Path Ahead

1. 'World leaders adopt Sustainable Development Goals', www.undp.org/2015/09/25.

2. World Bank Group, 2019, http://data.worldbank.org/indicatr/gb.XPD.RSDV.GD.ZS; Economic Survey, 2018, as reported in the *Economic Times*, 29 January 2018; http://data.gov.in/keywords/rd-expenditure.

3. See https://www.csir.res.in/achivement/csir-pride/healthcare-drugs-pharmaceuticals.

4. Nayan Kalnad, Niranjan Bose and Hari Menon, 'India's Great Healthcare Challenge, And Opportunity', The Quint, https://fit.thequint.com/health-news/opportunity-in-indias-healthcare-challenge-2.

5. See https://home.kpmg/in/en/home/insights/2021/02/india-healthcare-sector-transformation-in-the-post-covid-19-era.html.

6. K. Singh, P. Ghosh and D. Talukdar, 2015, 'India Healthcare Roadmap for 2025', Bain & Co., /www.bain.com/insights/india-healthcare-roadmap-for-2025-brief, cited by Turner P., 2018 in 'Talent Management in Healthcare', Palgrave Macmillan, Cham. https://doi.org/10.1007/978-3-319-57888-0_10.

7. See timesofindia.indiatimes.com/city/Dehradun/aiims-rishikesh-and-iit-r-successfully.....articleshow/7653.

8. R.B. Singh. Agricultural transformation–the road to new India, NAAS, India, 2020.

9. Nayan Kalnad, Niranjan Bose and Hari Menon, 'India's Great Healthcare Challenge, And Opportunity', *Hindustan Times*, June 2017, http://www.hindustantimes.com/india-news/indias-great-healthcare-challenge-and-opportunity/story-eB691Pj889x57TYo8MobaM.html; cited by Thejeshwar et al., 2020, medRxiv preprint doi: https://doi.org/10.1101/2020.08.13.20174144.

10. Open Letter to the United Nations, G-20, and National governments on Covid-19 and Agriculture for Food and Nutrition Security, May 2020, https://wle.cgiar.org/cosai/

sites/default/files/Open%20letter%20on%20Agriculture%20
Food%20%26%20Nutrition%20Security.pdf.

11. A. Gulati, M. Ferroni and Y. Zhou, 'Supporting Indian
    Farms the Smart Way', *Academic Foundation*, pp. 456, 2018.

12. A. Banerjee and E. Duflo, *Poor Economics*, Penguin Books India,
    New Delhi, p. 444, 2011.

13. See https://www.education.gov.in/sites/upload_files/mhrd/files/
    NEP_Final_English_0.pdf.

14. India's Energy Storage Mission: A Make-in-India Opportunity
    for Globally Competitive Battery Manufacturing, NITI Aayog
    and Rocky Mountain Institute, 2017; http://www.rmi.org/
    IndiasEnergy-Storage-Mission.

15. See https://www.livemint.com/Leisure/VVbKzzMdEzNzpF3vlx6
    WGP/The-Good-Earth-journey.html

16. See www.wadhwaniai.org.

Conclusion

1.   See http://mashelkarfoundation.org/amiia-awards/savemom-
     winner-details-2020/.